The Cambridge Technical Series
General Editor: P. Abbott, B.A.

TECHNICAL HANDBOOK
OF
OILS, FATS AND WAXES

VOLUME II

Practical and Analytical

METHOD OF USE OF COLOURED DIAGRAM
(Refractive Indices)

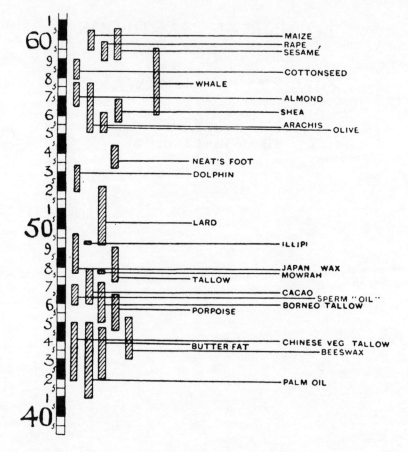

TECHNICAL HANDBOOK

OF

OILS, FATS AND WAXES

BY

PERCIVAL J. FRYER, F.I.C., F.C.S.

Chief Chemist and Director, Yalding Manufacturing Co. Ltd.,
Lecturer in Oils, Fats and Waxes at the Polytechnic, Regent Street, W.
Silver Medallist (1st Hons.), City and Guilds of London Institute

AND

FRANK E. WESTON, B.Sc. (1st Hons.), F.I.C.

Head of the Chemistry Department, the Polytechnic, Regent Street, W.
Author of *The Detection of Carbon Compounds* and *Elementary
Experimental Chemistry*

WITH 69 ILLUSTRATIONS

VOLUME II

Practical and Analytical

Cambridge:

at the University Press

1918

CAMBRIDGE
UNIVERSITY PRESS

University Printing House, Cambridge CB2 8BS, United Kingdom

Published in the United States of America by Cambridge University Press, New York

Cambridge University Press is part of the University of Cambridge.

It furthers the University's mission by disseminating knowledge in the pursuit of
education, learning and research at the highest international levels of excellence.

www.cambridge.org
Information on this title: www.cambridge.org/9781107660885

© Cambridge University Press 1918

First published 1918
First paperback edition 2014

A catalogue record for this publication is available from the British Library

ISBN 978-1-107-66088-5 Paperback

PREFACE

THE present volume of this treatise is concerned with the whole subject of the practical examination and analysis of the natural and hydrocarbon (mineral) oils, fats, and waxes. It is intended as a handbook of practical methods for the use of technical chemists and others concerned in the examination of oils and kindred substances, but the requirements of students of this subject have also been taken into account throughout the book.

Thus, the more important operations are illustrated by means of direct photographs in order that the methods of working, or the details of the apparatus may be clearly seen.

An endeavour has been made to continue the principle, aimed at in the first volume, of presenting the matter in a succinct and readily comprehensible form. To this end the subject matter has been carefully systematised throughout, and in the instructions for the performance of the various analytical operations the directions have been simplified by the use of concise language, and by subdivision into a series of short complete stages, or steps, each involving a single direction. The utility of this manner of presentation has been shown in the authors' own experience extending over a number of years.

It has also been our aim to eliminate, as far as possible, all untrustworthy details and methods of working. Some of the methods given are entirely original, as e. g. in the sections on Solubility, Viscosity, and Oxygen absorption. In many other cases modifications have been made in the procedure—the suggestions of the authors and others—in order to facilitate the work, shorten the time, or increase the accuracy of the results. In most cases the methods described have been carefully examined, and may be confidently accepted as reliable within the limits stated. It is well to remember in this connection, that many of the tests available are of quite an empirical character.

With regard to the arrangement of the book, the following brief description may be of service:

Section I is introductory, and outlines the main principles for

success in analytical work. Brief directions are given for the preparation of standard solutions and the use of simple apparatus is
explained by photographs for the student.

In Section II directions for sampling and for making preliminary
tests of the fatty material are given.

Section III deals with those tests of general application which
have a claim to be regarded as "standard analytical determinations."
To each of these is given the same indicating number which is
employed to designate it in the first volume, and this figure is adhered
to throughout the book to facilitate reference.

Section IV embodies those tests which concern particular oils or
groups of oils, and the authors have made a careful examination of
these, excluding those which were unreliable, and modifying the
procedure of others where this appeared desirable.

In Section V the examination of the mixed fatty acids from the
oils and fats has been considered, since in some important cases
(e.g. in the examination of soaps) the original oils are not available.
Tables are given of the more important analytical determinations,
and a revision of the refractive indices and of the iodine values has
been undertaken. The remaining portion deals with the alcohols,
with a full description of the methods for testing glycerin, both
crude and refined.

Section VI deals with the application of the foregoing methods to
the analysis of the hydrocarbon oils and waxes, together with the
special tests for these oils.

Rosin and turpentine, as being related in interest and also commercially with the natural oils and fats, are dealt with in Section VII
and methods of testing and of determining adulteration are given.

In Section VIII the authors have discussed the difficult question
of the interpretation of the analytical results obtained with commercial
samples. This includes the problem of the natural variation of oils
and fats, and of the modifications produced by various technical
operations. The question of the quantitative detection of adulteration is reviewed, and the oils are considered under the various
groups in which they are employed in commerce. A tabular state-
ment of the chief analytical data is included with each group of oils,
and practical examples of adulteration and its interpretation are
cited.

Section IX is intended primarily for the student. In it an attempt is made to give a systematic scheme of analysis for the identi-

fication of an oil, fat, or wax of assumed purity. The classification adopted in Volume I has been adhered to, and it is hoped that the scheme may be of value in familiarising the student—and others—with the distinguishing features and the class relationships of these compounds.

Section X is devoted to tables of strengths of solutions of various reagents, and to other data not included in the text of the book.

The authors wish to acknowledge the courtesy of Messrs Bolton and Revis and their publishers (Messrs Churchill) for permitting the use of the block of beef and lard crystals given on page 153. Messrs Hilger have been good enough to supply illustrations of their re-fractometers on pages 53 and 56, and have also given details of these instruments. We also desire to thank Mr G. B. Stokes and the publishers of the *Analyst* for permission to copy the illustration of the fat extractor on page 22.

PERCIVAL J. FRYER.

FRANK E. WESTON.

RAVENSCAR,
TONBRIDGE,
 and
THE POLYTECHNIC,
REGENT ST., W.
September 1918.

CONTENTS

SECTION IV

SECTION V

SECTION VII

SECTION VIII

SECTION IX

SECTION X

SECTION I

INTRODUCTION TO PRACTICAL WORK
FOR TECHNICAL STUDENTS

SECTION I

INTRODUCTION TO PRACTICAL WORK
FOR TECHNICAL STUDENTS[1]

§ 1. In order to obtain proficiency in analytical and experimental work, several factors are necessary for success. These may be classified as:

(*a*) Personal.

(*b*) Those relating to apparatus.

(*c*) Those concerning procedure.

§ 2. The **Personal factor** is of great importance in practical chemistry. It is essential for success to acquire the habits of *cleanliness* and *tidiness* in all work; attention should be paid to details, which *precludes hurry*, many operations requiring time for their performance if accurate results are to be obtained. Finally, the student must learn to *concentrate his mind* upon the work in hand and not attempt to do two or more operations at the same time. The result of this will be to generate that confidence in one's work which is the great secret of the successful worker.

§ 3. **Apparatus** plays a very important part in all analytical work. Generally speaking, there is a *right* way and *many wrong* ways of using apparatus. An endeavour should be made from the beginning to learn the correct method of using chemical apparatus, even the most simple. Measuring vessels should always be verified before they are used in carrying out any quantitative determinations, as much valuable time and annoyance will often thereby be saved. Complicated apparatus should be thoroughly examined and the *use of all parts ascertained* before its employment. Accurate results will only be obtained by repeated practice and careful attention to detail.

---

[1] This introductory chapter is intended to serve as a reminder to students of practical chemistry, and to some degree, perhaps, as a ready reference for qualified analytical and technical chemists. The successful performance of the practical analytical work in connection with the examination of oils and fats requires a high degree of training and experience in chemical theory and practice. The authors therefore strongly deprecate the too common practice of students taking up the study of applied chemistry, without qualifying in the general practical and theoretical branches of chemistry and physics.

§ 4. Even after one has become expert in manipulations, concordant results are sometimes unobtainable in certain determinations. This arises from some **inherent fault** in the method used, which *theoretically* may appear quite sound, but in practice reveals an error which frequently, when working under a given set of conditions, remains constant, and hence can be allowed for. Such methods however should not be employed if better ones can be devised.

§ 5. In the practical work described in this treatise the authors have endeavoured to give such instructions as will as far as possible ensure the acquisition of the **correct method of working**—many operations being illustrated, so that the best method of manipulation is readily seen. **Students** are directed also to the succeeding special section on the correct use of simple apparatus and the preparation of standard solutions for analysis.

Before commencing any test involving the use of even simple apparatus and reagents, it is always advisable to make sure that all the solutions are ready and accurately adjusted. All apparatus should be fixed up in readiness before the work is commenced, so that complete attention can be given to the manipulations.

To this end, in the chemical work described in this book, a list of the apparatus and special reagents necessary is given, unless these are likely to be immediately available.

§ 6. Many of the determinations carried out in the chemical examination of an oil, fat, or wax require the use of **standard solutions.**

A standard solution is one of known strength. The strength chosen in each case depends upon the use to which the solution is to be put, but it is always closely related to the *equivalent weight* of the compound being used.

§ 7. A solution of unit strength is known as a **normal solution,** and is one that contains one-equivalent-gramme weight of the compound in 1000 c.c. of the solution.

Consider the following equation :

$$NaOH + HCl = NaCl + H_2O$$
$$40 \qquad 36\cdot5 \quad 58\cdot5 \quad 18$$

It is seen that 40 grammes of NaOH neutralise 36·5 grammes of HCl, and since 36·5 grammes of HCl contain 1 gramme of replaceable H, 36·5 grammes of HCl is one-equivalent-gramme weight of hydrochloric acid, and 40 grammes of NaOH is one-equivalent-gramme weight of sodium hydrate. Hence a normal solution of hydrochloric acid contains 36·5 grammes of HCl per 1000 c.c.; and a normal solution of sodium hydrate contains 40 grammes of NaOH per 1000 c.c.

A normal solution is indicated by the letter N. placed before the chemical formula : thus **N. HCl** indicates a normal solution of hydrochloric acid.

§ 8. The following **table** gives a list of the compounds that will be required in the analysis of oils and fats, together with data concerning their equivalents, equivalent weights, etc.

Substance	Formula	Molecular weight	Equivalent	Equivalent weight	Weight of substance in 1000 c.c. of a N. solution
Hydrochloric acid... ...	HCl	36·5	1 molecule	36·5	36·5 grms
Sulphuric acid	H₂SO₄	98	½ ,,	49	49·0 ,,
Oxalic acid	H₂C₂O₄.2H₂O	126	½ ,,	63	63·0 ,,
Sodium hydrate	NaOH	40	1 molecule	40	40·0 grms
Potassium hydrate ...	KOH	56·1	1 ,,	56·1	56·1 ,,
Sodium carbonate... ...	Na₂CO₃	106	½ ,,	53	53·0 ,,
Sodium thiosulphate ...	Na₂S₂O₃.5H₂O	248·1	1 molecule	248·1	248·1 grms
Potassium dichromate ...	K₂Cr₂O₇	294·2	⅙ ,,	49·0	49·0 ,,
Potassium permanganate	KMnO₄	158·1	⅕ ,,	31·6	31·6 ,,
Iodine monochloride ...	ICl	162·4	½ molecule	81·2	81·2 grms
Iodine	I₂	253·8	½ ,,	126·9	126·9 ,,

In many cases these solutions will be too strong, and therefore solutions containing ½, ⅕, ¹⁄₁₀, etc. these weights are used. These solutions are termed semi-normal, fifth-normal, and decinormal respectively, and are represented by the signs N/2, N/5, N/10, etc., or ·5N, ·2N, ·1N, etc. If solutions stronger than Normal are required multiples of Normal solutions are used, as 2N, 2·5N, 3N, etc.

Fig. 1

PREPARATION OF STANDARD SOLUTIONS

§ 9. The simplest method of making a standard solution is to take a **known weight** of the substance required, dissolve in water, and make the solution up to the necessary volume by adding water to the required amount. This necessitates the following requirements:

(*a*) The use of a **pure** substance of *composition* corresponding to its *formula*.

(*b*) **Accuracy** in weighing.

(*c*) **Corrected** volumetric measuring apparatus.

(*d*) **Precision** in all the required manipulations.

Fig. 2

Condition (*a*) is fulfilled by only a VERY FEW of the compounds that are required, and hence the method is limited in application. Of those mentioned in the preceding table, only the following are suitable for direct weighing:

1. Pure anhydrous SODIUM CARBONATE Na_2CO_3.
2. Pure crystallised OXALIC ACID $H_2C_2O_4 . 2H_2O$.
3. Pure recrystallised POTASSIUM DICHROMATE $K_2Cr_2O_7$.
4. Pure recrystallised POTASSIUM PERMANGANATE $KMnO_4$.

Condition (*b*) depends upon the use of a good balance and **skill in weighing**. A short-beam balance capable of carrying a total load of 200 grms and weighing to ·0005 grm will be found suitable. The illustration (fig. 1) is a photograph of a Sartorius short-beam balance.

Condition (*c*) requires **accurately calibrated** measuring vessels ; these can be obtained quite easily at the chemical dealers. They should preferably be *checked* by actual weighing after additions of small volumes of pure recently-distilled water at standard temperature. An alternative method is comparison with a standardised apparatus. Fig. 2 is a photograph of the apparatus required. **a** represents a 250 c.c. *graduated cylinder* (50 c.c. and 1000 c.c. cylinders are also necessary), **b** is a 1000 c.c. *graduated flask* (50 c.c., 100 c.c., 250 c.c. flasks are also required), **c** shows *graduated pipettes*, 10, 20, and 50 c.c., and **d** represents a 50 c.c. *burette*.

Condition (*d*) is attained only by much practice and careful attention to detail.

§ 10. EXAMPLE OF DIRECT WEIGHING METHOD.—**Preparation of Decinormal (N/10) Sodium Carbonate.**

Apparatus required

balance sensitive to ½ milligramme	clay triangle
volumetric apparatus as in § 9	bunsen burner
desiccator	wash bottle
watch glass	funnel
crucible	small pestle and mortar
beakers	crucible tongs
tripod stand and wire gauze	250 c.c. Erlenmeyer flasks

Fig. 3

Materials

    [A] Pure anhydrous SODIUM CARBONATE.

    [B] DISTILLED WATER.

Procedure

1. Place about 10 grms of the well-powdered Na_2CO_3 [A] in a dry crucible and heat till bottom of crucible is dull red (fig. 3).
2. Transfer crucible to desiccator and allow to cool.
3. Take a clean dry watch glass, place on left pan of balance and accurately weigh it to $\frac{1}{2}$ milligramme (fig. 1).
4. Add 5·3 grms to the weights already taken and weigh out this amount of the dry Na_2CO_3 on the watch glass, taking from the centre of the crucible.
5. Transfer the weighed Na_2CO_3 to a clean 250 c.c. beaker, rinsing off all the powder into the beaker with the wash bottle (fig. 4).

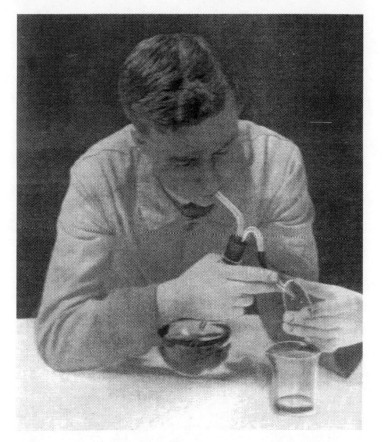

Fig. 4

6. Add to the beaker about 100 c.c. of distilled water [B] and gently heat till all is dissolved.

7. Take the 1000 c.c. graduated flask, rinse it out with distilled water, and insert a funnel in it.

Fig. 5

8. Carefully pour the Na_2CO_3 solution from the beaker into the flask through the funnel, using a stirring rod to direct the stream of liquid (fig. 5), and rinse out beaker by holding it mouth downwards over the funnel and directing a jet of water up into the beaker from the wash bottle (see fig. 4).

9. Rinse down the funnel inside and out with the wash bottle, remove the funnel, and add distilled water until the mark is almost reached.

10. Test temperature of solution; if below $15 \cdot 5°$ C., warm gently until this temperature is reached; if above, cool to the necessary degree by placing under a stream of water from the tap.

11. Add water to the mark on neck of flask; close with stopper, and mix contents by inverting flask several times, taking care that no leakage of the contents occurs. The solution so obtained is decinormal **(N/10) Sodium Carbonate.**

§ 11. N/10 solutions of oxalic acid, potassium dichromate, potassium permanganate and iodine are made in exactly the same manner, with the following exceptions:

The OXALIC ACID is not heated, but only well pressed between clean dry filter paper before weighing.

The POTASSIUM DICHROMATE is well powdered and heated in a crucible till it begins to fuse round the edges.

The POTASSIUM PERMANGANATE is well powdered and pressed between clean dry filter paper.

The IODINE is weighed in a stoppered weighing bottle, and is dissolved in potassium iodide solution or in alcohol.

In each case the WEIGHT TAKEN is $\frac{1}{10}$ part of the number of grammes for a normal solution, given in the foregoing table.

§ 12. Standard solutions of the **other compounds** mentioned in the table cannot be made by direct weighing, since the compound is

(1) known only in solution, or

(2) does not exist in a pure state.

Hence an **indirect method** has to be employed, involving the use of a standard solution made by direct weighing.

§ 13. EXAMPLE OF INDIRECT METHOD.—**Preparation of Decinormal (N/10) Hydrochloric Acid.**

Apparatus required

As before (§ 10) with addition of HYDROMETERS and 50 c.c. BURETTE IN STAND.

Materials

[A] Decinormal sodium carbonate.

[B] Pure concentrated hydrochloric acid.

[C] Methyl orange solution (1 per cent. in methylated spirit).

[D] Distilled water.

Procedure

1. Place 9 to 10 c.c. of the conc. HCl solution [B] in a 1000 c.c. cylinder and fill up with distilled water until the specific gravity is $1 \cdot 02$ (see p. 36). Mix by reversing several times. This gives a *slightly stronger* than N/10 solution.

2. The *exact strength* of this solution is now determined in the following manner:

(i) Take a dry 10 c.c. pipette, rinse with the N/10 Na_2CO_3 [A] (which should be at $15 \cdot 5°$ C. on testing with the thermometer: see § 10, above), and fill up to the mark with the same (fig. 6).

(ii) Run the measured 10 c.c. into a clean Erlenmeyer flask and add
2—3 drops of the methyl orange solution. This turns the solution light
yellow.

Fig. 6

(iii) Rinse out a 50 c.c. burette with the HCl solution (1) and then fill it to
the top division with the same solution. The solution should be preferably at standard temperature ($15 \cdot 5°$ C.), or in any case at the *same*
temperature as the N/10 Na_2CO_3.

(iv) Place flask (ii) under burette.   Note the reading of the burette (which should be preferably at zero), and then run acid into the flask 1 c.c. at a time (fig. 7), shaking with a rotary motion (produced by the wrist) after each addition, until the solution changes to a pink colour.   Note new reading of burette.   The difference between the two readings gives the volume of the acid used.

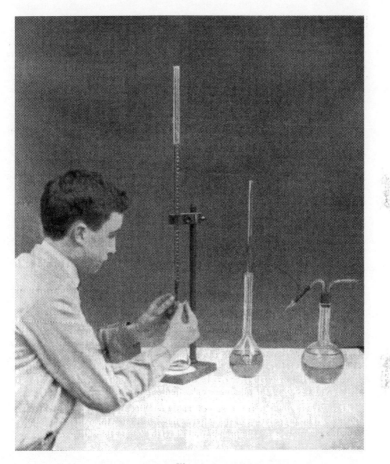

Fig. 7

(v) Empty the flask, well rinse with water, and measure out another 10 c.c. of N/10 Na$_2$CO$_3$, add 2—3 drops of methyl orange and again place under burette.

(vi) *Titrate* again with the HCl solution, but this time run slowly in the whole of the previous volume used less 1 c.c., shaking meanwhile with a rotary motion as before. Then proceed by adding 1 drop at a time from the burette, agitating after each addition, until a permanent pink coloration is obtained. Note the burette reading, and subtract from original reading, giving the volume of the acid used.

(vii) Repeat operation (vi) until two consecutive readings do not differ by more than ·05 c.c.

3. If the volume of the hydrochloric acid solution found to exactly neutralise the 10 c.c. $N/10$ Na_2CO_3 be represented by x c.c., then $x \times 100$ c.c. is the volume required for 1000 c.c. of $N/10$ Na_2CO_3. Hence, in order that the two solutions shall be equivalent, it is only necessary to take $x \times 100$ c.c. of the HCl solution and make up to 1000 c.c. by addition of distilled water. Assuming that the burette reading was 6·85 c.c. HCl solution to neutralise the 10 c.c. of $N/10$ Na_2CO_3, then proceed as follows:

(i)   Pour into 1000 c.c. graduated cylinder 6·85 × 100 = 685 c.c. of the HCl solution.

(ii) Transfer this to a clean 1000 c.c. stoppered flask, rinse out with distilled water into flask, and make up to the mark with distilled water at 15·5° C.

(iii) Insert stopper of flask and thoroughly mix contents by inverting several times.

4. Check this solution by titrating it against the $N/10$ Na_2CO_3. 10 c.c. of the latter should require exactly 10 c.c. of the acid. If the solution is still too strong, the calculated amount of water may be added to exactly standardise it. If too *weak*, a **factor** must be employed with the solution. Thus, if

10 c.c. of the Na_2CO_3 require 10·05 c.c. of the acid, then the factor is $\dfrac{10}{10\cdot05}$

or ·995, and whenever this solution is used for analysis, its readings must be reduced by multiplying by this factor: otherwise the readings would be too high.

5. Transfer to a good tight-stoppered bottle, and label immediately **N/10 HCl**, adding also the date of making, and initial the label. If a factor has been found necessary this should be stated on the label thus: "**factor** = ·995."

§ 14. Decinormal solutions of **sulphuric acid, sodium hydrate**, and **potassium hydrate** can be prepared in a similar manner. In the case of the last two, solutions of the approximate strength can be made up by dissolving 5 grms of NaOH and 6 grms of KOH respectively in 1 litre of water. The solutions are then titrated against $N/10$ HCl or $N/10$ H_2SO_4. In this case, the standard *acid* is placed in the burette, and 10 c.c. of the alkali are measured out.

§ 15. There remain the cases of **sodium thiosulphate** and **iodine monochloride**. The preparation of the latter is described later (p. 93). The following method is convenient for the preparation of $N/10$ sodium thiosulphate:

§ 16. Preparation of Decinormal (N/10) Sodium Thiosulphate.

The solution can be made by direct weighing (§ 10) if it is certain that the crystals are pure, but since they often contain occluded water, and commercial samples contain other impurities, it is not safe to rely upon this method.

Apparatus required
  As before (§§ 10, 13).

Materials

[A] DECINORMAL POTASSIUM DICHROMATE (§ 8) $N/10\ K_2Cr_2O_7$.

[B] POTASSIUM IODIDE SOLUTION.

Dissolve about 10 grms of KI in 100 c.c. water. It need only be roughly ten per cent.

[C] STARCH SOLUTION.

Grind up about 1 grm of starch into a paste with about 5 c.c. water, and pour this paste into about 100 c.c. of *boiling* water. Allow to cool before use.

Procedure

From the table (§ 8) it will be seen that 24·81 grms of the pure compound ($Na_2S_2O_3 . 5H_2O$) are required for 1 litre of $N/10$ solution. A solution of this approximate strength, but a little stronger, is therefore first made.

1. Weigh out, say, 26 grms of the crystals, dissolve in water and make up to 1000 c.c. in a graduated cylinder.

2. Fill a 50 c.c. burette with the solution.

3. Measure out with a pipette (fig. 6) 10 c.c. of $N/10\ K_2Cr_2O_7$ into an Erlenmeyer flask, add about 5 c.c. of the KI solution and 1 c.c. dilute[1] HCl. A reddish brown solution is obtained (due to liberated iodine).

4. Now run in slowly with constant agitation the $Na_2S_2O_3$ solution until the brown colour *nearly* disappears.

5. Add about 1 c.c. of starch solution, when solution will turn a dark olive colour. [If the blue-black colour does not appear, too much thiosulphate has been added and the titration must be begun again.]

6. Run in the thiosulphate solution again carefully **drop by drop** with constant agitation until the deep blue changes to a very pale green colour, and note number of c.c. of thiosulphate used.

7. Repeat until exact relation between two solutions is found.

8. Dilute the thiosulphate solution as described[2] in § 13, 3, i and ii, and check it by careful titration. If it is too weak, determine the factor as previously described, and record this on the label of the bottle.

---

[1] Bench HCl reagent is suitable, and is about 2N.

[2] Thus, if 10 c.c. $N/10\ K_2Cr_2O_7$ solution required 9·5 c.c. of $Na_2S_2O_3$ solution, then 1000 c.c. of $K_2Cr_2O_7$ will require $100 \times 9·5 = 950$ c.c. of $Na_2S_2O_3$ solution. Place 950 c c. of this last in a 1000 c.c. flask, and make up to the mark with distilled water. All solutions must be at $15·5°$ C.

SECTION II

SAMPLING AND PRELIMINARY TESTS

SECTION II

SAMPLING AND PRELIMINARY TESTS

§ 1. In the examination of materials supplied in bulk, it is essential in the first place that a sample is taken which **correctly represents** the total consignment. It frequently happens that a consignment of material comprises more than one manufactured charge, in the case of a factory product, or in other cases, that it is derived from various sources. Further, a number of packages may have been filled from a tank in which the liquid has settled, in which case those first filled may contain a higher proportion of impurities—such as e.g. in the case of oil; water, residues ("foots"), etc.—than the others.

The ideal method of sampling would be to empty all the packages into an effective mixing vessel, and obtain a thorough intermixture. This is termed "*bulking*," and in many cases, especially where uniformity in the manufactured product is desired, is actually carried out in the process of handling.

Failing this, the only other reliable method of obtaining a correct sample of a consignment is to get a representative sample of each package, together with the net weight of each lot, thoroughly mix each sample, and bulk them in the proportion of the net weights obtained. As the matter of accurate sampling is of great importance, it will be well to consider it in some detail.

Liquids

§ 2. A commonly employed method of sampling liquids is to take a piece of stout glass tubing of about three-eighths to one-half inch bore and to lower this slowly through the liquid. The ball of the thumb is then placed over the end of the tube and the column of liquid is withdrawn.

The drawback to this method is that, with very viscous liquids, such as castor oil, cylinder oils, or glycerin, a fair sample is not obtained, since the tube fills too slowly. Also it is necessary to take a correct proportion of the sediment which frequently collects at the bottom of the barrel, drum, or other package. To overcome this, the use of a specially designed **"sampler"** is advisable, such as the one shown in the figure. This consists of two metal tubes, one fitting closely within the other.

A number of slots are cut out in each tube, which exactly correspond when
the sampler is opened, enabling a com-
plete section of the liquid to be taken
almost instantaneously. There are also
openings cut in the bottom of the sampler
which admit the correct proportion of
the sediment, if any, at the bottom of
the package. The construction of the instrument is such, that by a simple action
of turning the handle provided at the
top all the slots can be opened simul-
taneously, and a pointer is arranged
which indicates when the sampler is
open or closed.

In using the sampler, it is introduced
into the barrel, drum, etc., with the
openings closed, and when it has touched
the bottom the slots are opened for a
second or two and then closed and the
sampler withdrawn. The charge is run
out by opening the slots again.

In the case of small packages, such
as cans and bottles, it is best to stir or
shake the contents thoroughly, and then
withdraw by means of a tube.

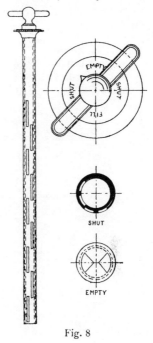

Fig. 8

Solids

§ 3. In the case of a soft solid, such as a fat, a cylindrical section of
the material is taken by means of an *auger*. It is advisable to take
several sections in different directions through the bung-hole or other
opening of the package. Each package is then weighed, the specimens
from each package softened, or melted at a low heat, and well mixed
together in proportion to the net weights so obtained. This method
generally suffices, but in extreme cases it may be advisable to take out
the head of the cask or barrel, and take samples lengthways as well.

Some waxes, which are too hard to admit of this treatment, are best
sampled by taking lumps from different parts of the sack or other con-
tainer, melting at a low heat, and mixing in the proportions of the net
weights. Or, in the case of brittle waxes, the samples may be powdered
and mixed.

Preliminary Tests

Water

§ 4. Water in oils, fats, waxes and rosin is indicated by the crackling
noise produced on heating a small quantity over a naked flame. If the
heating be done in a test-tube, there may be no crackling, but frothing

then occurs, and drops of water or a mist appears on the cool, upper parts of the tube. The estimation of water is carried out in the following manner:

§ 5. Oils

The water in non-drying oils may be estimated by the method given for fats. Drying and semi-drying oils are best treated as follows:

A small wide-mouthed flask **A** is fitted with a cork, and with two

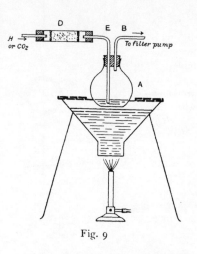

tubes as shown. It is then tared with the cork and tubes, about 10 grms of oil introduced, and the whole re-weighed accurately. A calcium chloride drying tube is placed at the end of tube **E**. It is placed on a water-bath, and a current of hydrogen or carbon dioxide drawn through the heated oil until the weight is constant.

In cases where water occurs in oils containing **volatile** constituents, such as crude petroleum, the water may be estimated by adding a volatile oil, such as xylene, boiling over 100° C., to a weighed amount

Fig. 9

of the sample, and then distilling until the distillate is nearly equal in volume to the added xylene, the distillate being collected in a graduated cylinder. The water appears underneath the latter in the receiver, and its volume may be directly read off.

Another method in the case of oils containing easily volatile acids is to spread about 5 grms of the oil over some freshly ignited sand in a basin, and place in a desiccator[1] over conc. sulphuric acid until no further loss in weight occurs.

§ 6. Fats and Waxes

The following method is suitable unless volatile matters are present, such as volatile fatty acids, traces of solvents, etc. In such cases the preceding method should be employed.

Accurately weigh into a tared dish about 5 grms of the sample and heat at 105° C. to constant weight. Where frothing is likely to occur, as in the case of some waxes (e.g. Carnaüba), a small stirring rod may be also weighed with the substance, and the fat or wax stirred from time to time.

---

[1] Preferably evacuated. Bolton and Revis use a crimped coil of filter paper, soaked in the oil and dried in a vacuum desiccator. *C. N.* Apr. 26, 1918.

In the case of some fats, or lubricating greases, which contain soaps (e.g. bone fat containing lime soaps), it will not be possible to remove the last trace of water. In such cases the amount of the water is best obtained by difference after estimating the fatty matter present and the non-fatty impurities (see below).

Foreign Substances

§ 7. Under this term are comprised cellular tissue from the parent substance, dirt, and other SUSPENDED MATTERS, inadvertently or fraudulently added. These are determined by the method (A) below.

WATER SOLUBLE SUBSTANCES may be removed and estimated by proceeding as under (B).

Naphtha, or other VOLATILE SUBSTANCES are determined by method (C).

(A) SUSPENDED MATTERS

10—20 grms of the dried oil, fat, or wax are dissolved in 100 c.c. of chloroform, and the solution poured on to a tared filter[1], any residue being washed with fresh solvent until the filtrate gives no residue on allowing a drop to evaporate on a clean watch glass.

The filter paper is dried at 100° C. and weighed. This gives the total suspended matter.

It should then be incinerated, and the ash examined as described in § 8. The difference is the amount of organic matter present.

(B) WATER SOLUBLE SUBSTANCES

About 50 grms of the oil, fat, or wax should be melted in a beaker with boiling water, and the whole poured into a separating funnel, shaken well, and allowed to separate.

The aqueous liquid may be tested for acid or alkali, and the amount, if present, determined by titration.

Evaporation of the liquid will give the total water soluble substances present, and the residue may be tested for metallic salts, soap, and starch by the usual methods.

(C) VOLATILE SUBSTANCES (except mineral oils)

These are determined by distilling 50 grms of the *undried* substance in a current of steam. The oil etc. is then dried at 100° C., the loss representing the moisture and volatile matter. The former is determined independently by the method given in § 5, and the difference gives volatile matters. The distillate may be examined for mineral naphtha, carbon disulphide, or other solvents, and for volatile fatty acids, ethereal oils, etc.

Mineral oils are dealt with on p. 198.

---

[1] The solution must be filtered hot in case of waxes, or an extraction apparatus may be employed (see p. 21).

§ 8. For the detection of **metals**, the oil etc. is shaken with hot dilute nitric acid in a separating funnel, and the metal tested for in the usual manner (for methods of recognising and estimating the metals[1] the student is referred to text-books on Inorganic Analysis).

The alkali metals and soaps are obtained in the test for water soluble substances (§ 7).

§ 9. **Sulphur, Nitrogen**, and **the Halogens** (chlorine, etc.) are tested for as follows:

Melt a pellet of sodium in a test-tube—supported vertically in a stand— and then allow about 1 c.c. of the oil to drop directly on to the sodium, drop by drop: then heat the test-tube to redness; allow to partially cool and cautiously plunge hot end of test-tube into 15 c.c. of water contained in a basin; well boil and filter. Test portions of filtrate as follows.

1. Rub a little on to a clean silver coin—a *black stain* indicates presence of *sulphur*.
2. To 5 c.c. of filtrate add a few drops of a solution of sodium nitro-prusside—a *violet coloration* indicates presence of **sulphur**.
3. To 5 c.c. of filtrate add 1 c.c. of a saturated solution of ferrous sulphate, a few drops of sodium hydrate solution, boil, add one or two drops of ferric chloride solution, and then concentrated hydro-chloric acid till solution is acid—a *blue coloration* or *blue precipitate* (Prussian blue) indicates presence of **nitrogen**.
4. To 5 c.c. of filtrate add dilute nitric acid till acid (well boil if nitrogen has been proved to be present) and then one to two c.c. of silver nitrate solution—a *white*, or *yellowish-white curdy precipitate* indicates presence of a **halogen** (chlorine, bromine, iodine[2]).

The above substances are estimated by the usual methods[3].

Determination of Fatty matter

§ 10. For the determination of total fatty matter in a substance a form of extracting apparatus is employed. If the substance is liquid, it is placed upon an absorbent material, such as sand or blotting paper, and if much water is present, dried first. Seeds are finely ground or shredded, taking care that none of the oil is lost by exudation whilst grinding. The substance is then placed in a fat-free thimble of absorbent paper, or a roll of filter paper formed into a pocket, and this placed in the extractor. The oldest form of extractor and the one frequently employed is the "SOXHLET." The authors have found that it is preferable to use an extractor in which the vapour of the solvent is allowed to act on the thimble. They have found the two forms described quite satisfactory.

---

[1] For the detection of nickel in hardened fats see p. 157.
[2] For further details of these tests consult *Detection of Carbon Compounds*, 3rd edn., Weston.
[3] See Clowes and Coleman, *Quantitative Analysis*.

§ 11. BOLTON AND REVIS fat extractor. The designers give the following directions:

The fitting up of the extractor is seen by the figures. The outer tube is indented about 2 inches above the neck in order to carry the extractor proper, which is lightly plugged at its lower end with cotton wool, and has a disc of filter paper resting on the wool to prevent the material to be extracted passing into the wool. The lower end of the extractor just enters a folded filter placed in the neck of the outer tube. The quantity of solvent should not be enough to fill the extraction tube when working.

The flask should be sunk to the shoulder in a water-bath which may be safely and efficiently heated by means of a 16 c.p. electric lamp, partially submerged in the water, and as rapid a rate of flow is maintained as is consistent with safety.

As a general rule from a half to two hours is sufficient for the first extraction, after which the extractor is removed and the solvent allowed to spontaneously evaporate in a *moderately* warm place

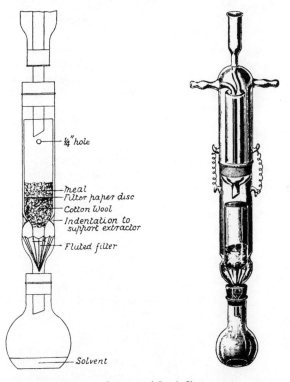

Fig. 1c.   Bolton and Revis Extractor

(too rapid heating may cause ejection of the contents). When the solvent has evaporated the tube is placed in a water oven for about an hour, the contents having been first distributed over the walls of the tube by gentle tapping.

After this, the contents of the tube are transferred to a mortar together with 1—2 grms of clean dry sand (passing 60 mesh) and ground as finely as possible, the contents of the mortar being then carefully brushed back into the extractor and the mortar washed out with the solvent.

Extraction is then continued for an hour to an hour and a half, after which the flask is detached and the solvent distilled off with the flask in an inclined position (so as to prevent splashing of the fat on to the delivery tube), the flask being then placed in a water oven and dried to constant weight. The flask may with advantage be inclined sideways and turned from time to time.

If heating be not unduly prolonged, there is little danger of appreciable increase of weight due to oxidation, but a temperature of 105° C. should not be exceeded.

A few oils[1], owing to the rapidity of their oxidation in air, must be dried in a current of indifferent gas, such as CO_2 or hydrogen.

§ 12.   Another simple but convenient and efficient form of extractor is that designed by *G. B. Stokes*[2]. It is especially suitable for waxes, and for materials requiring a hot extraction. The apparatus is shown in the figure. It consists of a flask with an extra wide neck, a piece of wire, bent round as shown, so as to enclose a thimble of fat-free paper. The wire passes through a hole in the cork, enabling the thimble to be raised or lowered at will.

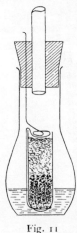

Into the thimble the substance to be extracted is placed, and above it a layer of fat-free cotton wool to prevent any of the substance floating over. It is placed within the wire frame, and at the commencement the thimble is immersed in the solvent.

After a suitable time the thimble is raised out of the liquid by means of the wire, without disconnecting the flask, and as the hot vapours circulate round the thimble, this is thus washed inside and out by the solvent.

When the extraction is complete the solvent is evaporated, and the flask and its contents weighed.

The authors have not found the **Soxhlet** form of extractor convenient for fat extraction, since the material is not exposed to the vapours of the solvent and the action of the siphon is normally too rapid to empty the thimble of its contained liquid, thus rendering the extraction unnecessarily slow.

Fig. 11

---

[1] E.g. the oil described by Bolton and Revis, *Analyst*.   [2] *Analyst*, 1914, **39**, 295.

If the improved forms of extractors are not available, the Soxhlet may be employed, but it is well before stopping the extraction to test the solvent in order to see that it is free from fat. This may be done by removing a few c.c. by means of a pipette just before the siphon commences to act and evaporating it on a clean watch glass, when the surface of the glass should remain perfectly clean.

§ 13. SOXHLET FAT EXTRACTOR

The photograph shows the form of the extractor. It is connected with the flask (**d**) and condenser (**e**), the former containing the solvent and heated on the water-bath. The vapour of the solvent passes up the tube (**a**) connecting with the upper part of the extractor, and the condensed solvent falls continuously on the material to be extracted contained in a thimble of filtering material (**b**). When the liquid reaches a given level, the siphon (**c**) comes into action, and the solvent runs back into the flask, where it is continuously evaporated. At the conclusion of the extraction, the flask (**d**) is removed, and the solvent evaporated. The residual oil is then dried at 100° C. to constant weight.

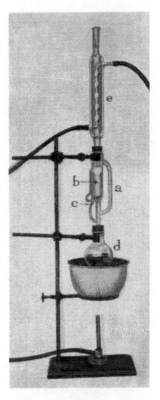

§ 14. As regards the choice of a suitable **solvent** for fat extraction, petroleum ether is preferable in most cases, as it extracts less non-fatty matter than methylated ether. It should however be carefully fractionated, using a good fractionating column, and for ordinary work all fractions above 80° C. rejected, while for special work (particularly in extracting drying oils when prolonged heating is undesirable) a spirit boiling below 40° C. is preferable.

Fig. 12. Soxhlet Extractor

Preparation of the Sample for Analysis

§ 15. The oil, fat, or wax should be melted at a gentle heat, and if then perfectly clear and bright, needs no preparatory treatment, being suitable for any of the tests described in the following sections.

If the fat etc. is not clear, it should be filtered (a hot-water funnel is necessary in the case of fats and waxes) through thick soft filter paper, previously dried.

The oil etc. obtained after estimation of the water by drying (see above) is suitable for testing if clear; that obtained by extraction with a solvent also requires no preparation, provided it has been dried to constant weight, and is therefore free from solvent and moisture.

SECTION III

PRACTICAL METHODS FOR THE STANDARD ANALYTICAL DETERMINATIONS

(1) Specific Gravity at 15·5° C. $\left[\dfrac{15\cdot5°}{15\cdot5°}\,\text{C.}\right]$

(2) Melting Point °C.

(3) Solidifying Point of mixed fatty acids (Titer test)

(4) Refractive Power

    (*a*) Dispersive Power

(5) Viscosity

(6) Solubility. "True Valenta"

(7) Rotatory Power in Polarimeter

(8) Iodine Value (and its variants):

    (*a*) Bromine Thermal Test;
    (*b*) Maumené Test;
    (*c*) Livache

(9) Elaïdin Test

(10) Saponification Value

(11) Insoluble Bromide Value of mixed fatty acids

(12) Reichert-Meissl and (*a*) Polenske Values

(13) Acetyl Value

(14) Acid Value

(15) Unsaponifiable matter per cent.

SPECIFIC GRAVITY   (1)

§ 1. Definition
The *specific gravity* is the weight of any volume of a substance compared with the weight of an equal volume of pure water at a standard temperature, e.g. 15·5° C.

§ 2. General Remarks
Many methods have been devised of determining the specific gravities of oils, fats and waxes, but for general analytical and technical work some of these are too lengthy and give a higher degree of accuracy than is required. It will be found that the following methods meet all cases arising in practice:

(a) THE SPECIFIC GRAVITY BOTTLE.
(b) THE SPRENGEL TUBE.
(c) THE HYDROSTATIC ("WESTPHAL") BALANCE.
(d) FLOTATION IN ALCOHOL.
(e) HYDROMETERS.

As regards *temperature*, the observations recorded in Vol. I of this book have all been obtained by adhering to one standard temperature for all determinations, e.g. **15·5° C.**[1], i.e. the oil at this temperature is compared with water at 15·5° C. This is expressed thus, S.G. $\frac{15\cdot5°}{15\cdot5°}$ C. The determinations of the mixed fatty acids are taken at the temperature of boiling water, i.e. approximately $\frac{99°}{15\cdot5°}$ C. [Where no mention of the temperature of the water is made, it is understood to be taken at 15·5° C.]

Determinations of the S.G. of oils and fats have, in the past, been made at very diverse temperatures. It is very desirable to adhere to the above standards for future observations, so that strictly comparable results may be obtained.

(a) THE SPECIFIC GRAVITY BOTTLE

§ 3. Outline of Method
The bottle is first weighed empty, and again when filled with the oil. The net weight of the oil is then divided by the known weight of water which the bottle will contain.

---

[1] For scientific purposes, comparison is usually made with water at 4° C., when it is at the maximum density.

§ 4. **Remarks**

A bottle of 50 grms of water capacity at 15·5° C. is the most suitable to use. An accuracy of ± ·0001 in the S.G. can be attained. The capacity of the bottle should be verified when first used by means of recently boiled distilled water, and tested periodically. A correction, if necessary, must be made.

§ 5. **Procedure**

1. Dry and clean the bottle and stopper (the latter has a small hole pierced through it).

   The drying may be rapidly performed by rinsing with (i) small quantities

Fig. 13

of dry alcohol, followed by (ii) small quantities of dry ether, draining the latter from the bottle, and blowing *dry* air into it.

2. Insert the stopper and weigh bottle accurately to four places of decimals. Note weight (p).

3. If the oil is not perfectly clear there may be water present in it. In this case, warm until the oil is about 75° C., and, if still not clear, filter through a dry filter paper.

4. Cool the oil to 15·5° C., or if below this, warm with the hand to this temperature.

5. Fill the bottle with the oil.

The diagram (fig. 13) explains the best method. The oil is poured slowly and carefully into the bottle which is meanwhile held in a slanting position so that air-bubbles are not taken in with the oil. When full, stand on the bench and allow any air-bubbles formed to rise to the surface. Remove the froth produced by these and carefully insert stopper, squeezing excess of oil through the capillary tube in the stopper.

6. Wipe off excess of oil with a soft cloth, first from the stopper and then from the rest of bottle, *holding the latter by the neck*[1].

7. Carefully weigh bottle + oil, and note weight (q).

8. Empty oil back into oil-bottle, allowing it to flow over the bulb of a thermometer so as to ascertain exact temperature.

§ 6. **Calculation of Result**

The weight of oil at $x°$ C. $= q - p$,

$$\text{S.G. } \frac{x°}{15\cdot5°} \text{ C.} = \frac{\text{weight of oil}}{\text{weight of equal volume water}} = \frac{\text{weight of oil}}{50^*},$$

$$\text{S.G. } \frac{15\cdot5°}{15\cdot5°} \text{ C.} = \frac{\text{weight of oil}}{50} \pm (\cdot00064 t°),$$

where $t°$ is the difference between $x°$ and 15·5° C., and is to be subtracted when $x°$ is below 15·5° C. and added when $x°$ is above 15·5° C.

The correction given (·00064) is an approximation from determinations of the coefficients of expansion of various oils[2]. As oils do not expand quite equally, it is not strictly accurate in all cases. Further the same oil does not expand equally between different ranges of temperature. For small differences of temperature involving small corrections it is however sufficiently accurate.

Note. If several determinations are to be made, it is only necessary to rinse the bottle out twice with the new sample of oil, drain, and then fill as before.

Students should always leave bottles quite clean when finished with by washing with hot soft soap solution, and rinsing with clean water.

---

[1] The warmth of the hand would expand the oil which would issue from the stopper and might be lost.

[2] Allen.

[*] Correction is made for capacity of bottle if necessary.

(b) **THE SPRENGEL TUBE**

This method should be used:

A. When a small amount only of oil is available, viz. 25 c.c. or less.

B. In the case of viscous oils and fats which require a more or less elevated temperature to render them fluid.

C. When the specific gravity is required to be taken (as in the case of mixed fatty acids) at a higher temperature than 15·5° C. This should be at 99° C. (at temperature of boiling water, see § 2 above).

Fig. 14. Sprengel Tube

§ 7. Description

The apparatus (see fig. 14) consists of a U-tube ending in two horizontal capillary tubes **a** and **b**, the former of which has a finer bore than the latter, which is provided with a mark a short distance from the end. The two extremities are provided with ground-on glass caps.

§ 8. Outline of Method

A. The tube is filled, by suction, with the oil, weighed, and the net weight of the oil divided by the weight of an equal volume of water.

B. The oil is heated until quite fluid, the tube filled as before, and then laid in a flat dish, covered with oil, and the whole allowed to cool to standard temperature before weighing.

C. The tube is filled and then immersed in hot water at the required temperature.

§ 9. Procedure

A. 1. Clean and dry the Sprengel tube by the method described in § 5 (1).

Fig. 15

2. Attach a piece of rubber tubing to **a** (fig. 14), insert **b** into liquid required and suck rubber tube until the Sprengel is full (fig. 15).

3. Remove excess of oil by applying a piece of blotting paper at end **a** until oil stands exactly at mark **b** (fig. 14). If too much oil is taken out, bring a drop of the oil at the end of a glass rod to end **a** when a little oil will enter.

4. Adjust glass caps over the ends of tube, support with wire loop on hook of balance, and weigh tube carefully to four places of decimals (q).

5. Subtract the previously ascertained weight of the tube from weight (q)=net weight of oil. Divide this by the ascertained capacity of water. This should be frequently checked by filling as described with recently boiled distilled water.

6. Ascertain temperature of oil as described in § 5 (8), and calculate s.G. as shown previously, § 6.

B. With viscous oils, or those containing deposited "stearine," the oil is heated to about 75° C., filled into tube as before described, and then laid in a flat "petri" dish filled with the heated oil. It is allowed to cool to 15·5° C., and is then taken out, wiped clean with a soft cloth, the oil adjusted as previously described, the caps placed on, the tube weighed, and the s.G calculated as before.

C. When the s.G. at elevated temperatures is required, the oil is heated on the water-bath until a few degrees under the final temperature. The Sprengel tube is then filled as before and placed in a water-bath heated to the necessary constant temperature (preferably 99° C., see above). The excess of oil exudes from the tube and, after about ¼ hour, the oil is adjusted to the required mark A 3, the tube dried and weighed, and the s.G. calculated as shown above[1], A 5 and § 6.

Students should leave Sprengel tubes free from oil. This may be best accomplished by blowing out oil through larger end, filling the tube with hot soft soap solution twice (by suction) and finally with warm water.

---

[1] A correction is necessary in very accurate work for the expansion of the glass.

(c) THE HYDROSTATIC (e.g. "WESTPHAL") BALANCE

§ 10. This instrument is extremely useful when such a degree of accuracy as that given by the S.G. bottle or the Sprengel tube is not required. Moreover, the determinations are rapidly carried out, saving much time. The method however requires *a larger quantity of oil* for the determination, and is most suitable for the less viscous oils and for general works' control.

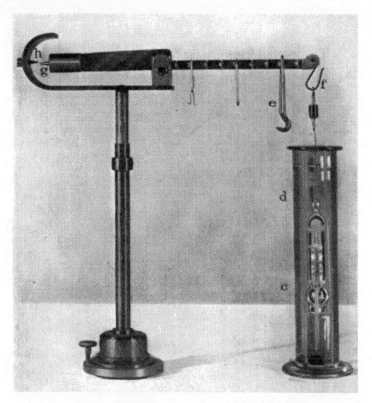

Fig. 16. Westphal Balance

§ 11. Description

A usual type of the instrument is shown in the photograph (fig. 16). This is the **"Westphal"** hydrostatic balance. It consists of a beam, resting on a knife edge, one end having a balancing weight and carrying a pointer, the other being provided with notches or pegs on which to suspend the weights ("riders"). This end carries the "plummet" (**c** in fig.), a small float weighted with mercury and usually containing a thermometer. This is suspended from a hook by a fine platinum wire, and the whole hangs from the end of the balance arm. The foot has a levelling screw for adjustment of the instrument.

§ 12. Outline of Method

The balance is first ADJUSTED by suspending the plummet **c** in a cylinder **d** of recently boiled distilled water at 15·5° C. The pointer **g** should now be exactly opposite the fixed indicator point **h**, when the largest rider **e** is placed on the hook **f**. If it is not, the screw in the foot must be turned so as to tilt the instrument until the two points are together.

The plummet is dried and the instrument is used in the same manner for the oil to be tested, but weights are placed on the beam until the points are in exact juxtaposition.

§ 13. Procedure

1. Adjust balance as described in § 12.
2. Wipe plummet dry, being careful that the fragile platinum wire is not bent or broken. Fill the cylinder with oil at the standard temperature, 15·5° C.
3. Hold plummet by brass hook at top, and dip into the cylinder of oil. Then lift out, and blow out the air-bubble which will have formed in the eye-hole from which the plummet is suspended.
4. Re-immerse in the oil, and adjust the height of the balance (by means of telescopic sliding pillar) until the plummet hangs about in the middle of the oil, and has plenty of clearance from the bottom.
5. See that the plummet is quite free from the sides of the cylinder. Then add riders until the points **g** and **h** are exactly opposite each other. Read off directly the specific gravity from the riders used, commencing with the largest, and ending with the smallest.

§ 14. Calculation of Result

No calculation is necessary, the instrument giving a *direct* reading. Thus :

    (*a*) The largest rider indicates from 1 to 0·1 S.G., according as position is 1—10 on beam.

    (*b*) The next largest rider indicates from 0·1 to 0·01 S.G., according as position is 1—10 on beam.

    (*c*) The next largest rider indicates from 0·01 to 0·001 S.G., according as position is 1—10 on beam.

    (*d*) The smallest rider indicates from 0·001 to 0·0001 S.G., according as position is 1—10 on beam.

      In the photograph (fig. 16) the S.G. is ·8522.

Students should cleanse plummet by dipping it into hot soft soap solution, rinse in water, and carefully dry with a soft cloth, avoiding injury to the fragile platinum wire. Avoid breakage by not holding the plummet suspended by the wire when cleaning.

(d) THE FLOTATION METHOD

This method is useful for the determination of the S.G. of solids, especially waxes. It may also be employed for some hard fats.

A slight modification of the following method may also be employed to ascertain the approximate S.G. of oils of which only a few drops are available for examination.

§ 15. Description

The principle made use of is that a solid will *just float* in a liquid of its own specific gravity. Hence a number of liquids of varying specific gravities will be required. The most convenient liquid is alcohol, and this is diluted with varying amounts of water. Since the specific gravities of the natural waxes lie between ·940 and ·999, it will be necessary to have a series of dilutions between these densities.

§ 16. Outline of Method

Small, smooth-surfaced pieces of the substance are prepared, free from occluded air-spaces, and placed in the varying solutions in order of densities, until a solution is found in which they neither sink nor rise, but just float. (Ideally they should remain suspended at all depths of the solution; in practice sufficient accuracy is obtained if the substance very slowly rises and remains *just floating.*)

§ 17. Procedure

1. Prepare six solutions of alcohol[1] and water by means of the Westphal balance, with densities ranging from ·940 to 1·00 (water only), differing by ·01, and keep in well-stoppered bottles. (Number them 1—6.)
2. Melt about 5 grms of the sample of wax in a small porcelain basin floating in a beaker containing boiling water. See that the liquid is free from bubbles, and quite clear. (If cloudy, filter through hot funnel and dry filter paper.) Allow to cool slowly.
3. When cold, cut out a small pellet of the wax, and to prevent the formation of air-bubbles, brush over with a wet camel hair brush.
4. Test the solutions, and see that they are *about* 15·5°C. Then place solution 1 in a tall-shaped beaker, and immerse pellet of wax by means of forceps.
5. If the wax sinks to the bottom, pour out solution 1 back into its bottle, fill with solution 2, and so on through the series until a solution is found in which the wax *floats.*
6. Cool or warm the solution until it is exactly 15·5°C. Fill a burette with alcohol, and add this drop by drop to the solution in the

[1] Ordinary methylated spirit will serve for any but special cases.

beaker until the pellet of wax *just* sinks, meanwhile mixing very gently by means of a stirring rod to prevent air-bubbles forming round the wax. If the amount of alcohol is exceeded, a drop or two of water may be added, and the end point again obtained.

7. Ascertain carefully the S.G. of this solution by means of the Westphal, or the Bottle method. This gives the S.G. of the wax.

8. Confirm this result by placing two more pellets of the wax (obtained as in (3) above) in this final solution. If the wax floats, test by adding a drop of alcohol to the solution, when it should slowly sink. If the pellet sinks, add water until it just rises, and ascertain S.G. as before. Obtain two consecutive concordant results.

Students should return all the solutions to their correct bottles, except the final one, which should be thrown away, and the bottle left empty.

(e) HYDROMETERS

§ 18. The use of a **hydrometer** affords the quickest method of ascertaining the S.G. of a liquid. It is, however, the least accurate of those which have been described, and this is because there is

(i) No direct weighing.

(ii) A liability to personal error in taking readings (see below).

(iii) A possibility of an error in the calibration of the instrument.

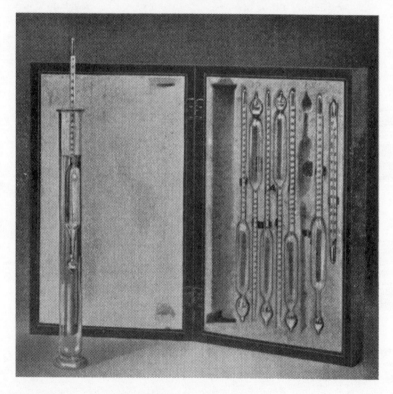

Fig. 17. Hydrometers

It is however an exceedingly useful method for general technical and works' practice, and is so simple, that the untrained workman can be trusted to use it correctly.

§ 19. Description

A hydrometer is a float, consisting usually of a cylindrical glass tube weighted at the bottom end with either mercury or shot, and graduated on the stem with a scale which either gives the specific gravities direct, or on some arbitrary scale of numbers (as e.g. the *Twaddell* or *Baumé*[1]) which are convertible into specific gravities by calculation. These latter are usually employed in works, for simplicity's sake, but it is best to use the direct reading instrument in the laboratory.

§ 20. **Outline of Method**

The hydrometer is placed in a sufficient bulk of the liquid, and since it becomes more or less immersed according to the density, the specific gravity can be directly read off on the scale placed inside the stem.

§ 21. **Procedure**

1. Clean and dry the glass cylinder for holding the oil to be tested, and fill to the necessary depth with the sample.
2. *Carefully* insert the hydrometer, avoiding violent contact of the small terminal bulb with the bottom of the cylinder.
3. See that the hydrometer floats *freely*. If it floats with the scale completely out of the liquid, or if it rests on the bottom of the cylinder, it must be removed, and a heavier or lighter one inserted.
4. Read the scale at the point on the stem of the hydrometer LEVEL WITH THE SURFACE OF THE LIQUID : to do this place the eye on a level with the surface (see fig. 17).
5. Ascertain the temperature of the oil, and make a correction as before ($\pm$ ·00064 per degree) so as to obtain result at 15·5° C.
6. Pour out the oil, and, if fresh samples are to be tested, rinse out the cylinder *twice* with the new sample before filling up.

Students should carefully clean hydrometer and cylinder with warm soft soap solution, and dry with a soft cloth when work is finished.

[1] Tables for the conversion of Twaddell and Baumé degrees into specific gravities are given at the end of Vol. I.

MELTING POINT (°C.)   ②

Definition

The melting point of a solid substance is the temperature at which it changes into its liquid state.

Remarks

All CHEMICALLY PURE substances have well-defined melting points which can be ascertained to a fraction of a degree, the substance changing into its liquid form at a constant temperature. Fats, and oils solidified by cooling, do not give well-defined melting points, since they are mixtures of glycerides, and contain also small quantities of other substances. Further, the phenomenon of a *double melting point* is exhibited by fats, the lower M.P. being given by the non-crystalline form[1]. When fats are gradually heated they begin to melt at a lower temperature than is required to convert the whole into a clear liquid. Thus two points are ascertainable, viz.:

(*a*) Point of *incipient fusion*.

(*b*) Point of *complete fusion*.

The point of *incipient fusion* is noted when the fat is seen to begin to lose its opacity and become semi-transparent. That of *complete fusion* is the point when the fat is observed to become perfectly transparent, and to contain no opaque particles. This latter point is usually given as the melting point.

Description

The standard method for melting point determinations is known as the **capillary method**. As in the case of fats, it is often a matter of great difficulty to observe accurately the stages of conversion into liquid form in so small a space as a capillary tube, an alternative method is given which has been recently proposed[2], in which a thin film of fat of comparatively large area is employed[3].

(a) CAPILLARY METHOD

Outline of Method

The fat or wax is melted, and introduced into a fine capillary tube, this is attached to an accurate thermometer near the bulb, and the temperature

---

[1] See Vol. I, page 12.

[2] *Giorn. di Farm. e di Chim.* 1916, **4**, 151—153.

[3] Such modifications as "*the dropping point*" (Ubbelohde) do not appear to the authors to offer any advantage over the direct observation methods.

is gradually raised until the fat melts. If an oil, this is solidified in a freezing mixture, and the determination carried out on the frozen material.

Apparatus required

[A]   A piece of soft glass tubing about 4 mm. bore ($\frac{3}{16}$ inch).

[B]   An accurate thermometer, preferably reading to $\frac{1}{10}$ degree.
For some oils, one with a range to $-30°$ C. will be necessary.

[C]   Two small rubber bands cut from a piece of $\frac{1}{4}''$ bore rubber tubing.

[D]   500 c.c. beaker, provided with glass-rod stirrer.

[E]   Small crucible.

[F]   Tripod stand and wire gauze.

[G]   Clamp for thermometer.

Procedure

1.   Heat tube [A] in the bunsen flame till soft, rotating it all the time to ensure even heating. Remove from the flame and draw out slowly to the required thickness, viz. about ·025 inch bore and ·003 inch thickness in the walls.

2.   Melt the fat or wax in a small basin, using only gentle heat, and breaking off one large end of glass tube, fill about 4 inches of the capillary tube with the fat or wax by suction.

3.   As soon as the fat is set, break the tube into pieces of about 1 inch long, and place these on ice for at least 30 minutes (or leave at normal temperature for 24 hours). (In the case of an oil, cool in a freezing mixture to at least 5° below M.P.)

4.   Attach a capillary tube **a** to a thermometer **b** by a thin rubber band **c** so that the fat is opposite the bulb of the thermometer (fig. 18).

5.   Suspend thermometer in the 500 c.c. beaker, filled preferably with recently boiled distilled water and placed on the tripod stand over the gauze. This is done by clamping the thermometer at its upper end. The bulb of the thermometer should be about $\frac{2}{3}$ up from the bottom of the beaker.

6.   Heat the water with a bunsen flame so that the rise in temperature is about 5° per minute, keeping meanwhile well stirred. Note temperature when fat becomes melted to a clear liquid. (In the case of an oil, place the thermometer in alcohol at the freezing temperature and allow the temperature to rise spontaneously.)

7.   Place fresh capillary tube in position, and fill the beaker with water again. This time heat at rate of 2° C. rise per minute until within 10° from the point previously observed. Then turn down flame and heat at not over $\frac{1}{2}°$ C. per minute, stirring well. Note carefully (preferably with a lens) the temperature at which the fat begins to appear translucent and shows a meniscus at top of tube,

and the temperature when complete transparence is obtained. These two points are recorded as the *incipient fusion* and *complete fusion* of the fat.

8. Repeat until two concordant readings are obtained.

Fig. 18

(b) PLATINUM LOOP METHOD

Outline of Method

A short piece of platinum wire is attached to a thermometer just over the bulb and this is made to form a loop in front of and close to the bulb. A film of fat or wax is formed in the loop, and temperature is noted at which the fat or wax becomes completely transparent, just before breaking up.

Apparatus required

As for method (*a*) but without capillary tube and with the addition of a pair of pliers and $1\frac{1}{2}$ inches of fine platinum wire.

Procedure

1. Bend the platinum wire about $\frac{1}{3}$ from one end and loop it round thermometer just over top of bulb (see fig. 19).
2. With the pliers, grasp the two ends at their junction **a** and carefully twist a few times, so as to give the wire a firm hold on the thermometer. Avoid too tight a twist as this might break wire or crack thermometer (fig. 19).

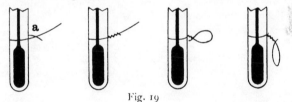

Fig. 19

3. Now bend the remaining wire into the form of a loop as in fig. 19.
4. Bend the loop down so that it is opposite and nearly touching bulb of thermometer (fig. 19).
5. Melt the fat or wax to be tested at a low temperature (preferably over steam) in a watch glass, place the loop so that it just dips below surface of melted liquid, and a film of fat or wax is formed in the loop (fig. 20).

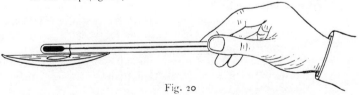

Fig. 20

6. Allow to cool on ice for 30 minutes or leave for at least 24 hours at normal temperature. Then ascertain melting point as previously described under method (*a*).

When work is finished students should carefully cleanse thermometer and loop by dipping into hot soft soap solution and rinsing in clean water. Leave loop attached to thermometer.

SOLIDIFYING POINT OF THE MIXED FATTY ACIDS
from an Oil or Fat (°C.)   ③
(TITER TEST)

Definition

The solidifying point is the temperature at which the liquid fatty substance changes into its solid state.

Remarks

It is difficult to ascertain accurately the solidifying point of an oil, a fat, or the fatty acids derived therefrom, by observing the temperature at which turbidity takes place. A much more reliable method is based on the fact that changing from a liquid to a solid state, bodies give up the LATENT HEAT OF FUSION ; and this is in most cases readily observable with a delicate thermometer. For this purpose, one graduated in $\frac{1}{10}$ degree C. is preferable. Fats and oils do not show this rise as clearly as the fatty acids derived therefrom ; and the test is therefore made with the latter. Useful information is however often obtained by *comparing the* S.P. *of the oils or fats with the* S.P. *given by the mixed fatty acids of the same.*

Outline of Method

The fatty acids are liberated from the oil with great care to avoid oxidation, and to obtain them free from moisture ; the solidifying point is then taken by *Dalican's* method, as under, the temperature of the thermometer being carefully observed until it remains stationary, and finally commences to rise. The **top point of the rise** is the figure required.

Apparatus required

[A]   A test-tube, 16 cm. long by 3·5 cm. wide.

[B]   A wide-mouthed bottle, 10 cm. wide by 13 cm. high, provided with a cork, having a central hole through which is inserted the test-tube, and which acts as an air-jacket for the latter.

[C]   An accurate thermometer, graduated in $\frac{1}{10}$°C., inserted into a cork which supports it so that the bulb of the thermometer is in the centre of the tube and about 3 cm. from bottom (see fig. 21).

[D]   A large (600 c.c.) beaker.

[E]   A water-bath.

[F]   Large funnel and thick dry filter paper.

[G]   A porcelain basin.

[H]   A desiccator.

[I]   A conical flask, capacity about 300 c.c.

Procedure. (If the fatty acids are already obtained, commence from (9).)

1. About 50 grms of the fat are poured into the flask and roughly checked on a balance.

2. About 15 grms of stick caustic soda are added, and 100 c.c. of 95 per cent. alcohol.

3. The fat is saponified, as described under "Saponification Value" (page 106).

4. The alcohol is evaporated off in the water-bath, and the soap dissolved in a little hot water, rinsed into the beaker, and made up to about 500 c.c. with hot water[1].

5. A few drops of methyl orange are added, and then sufficient hydrochloric acid (1 : 2 strength) to make the liquid *permanently* pink.

6. The beaker is heated on a water-bath (best immersed in the boiling water) until the fatty acid layer is PERFECTLY CLEAR.

7. The watery acid liquor is removed with a large pipette or siphoned off as completely as possible, and the fatty acids filtered through a perfectly dry, thick filter paper, into the porcelain basin, and the operation repeated if necessary until a *perfectly clear* filtrate results.

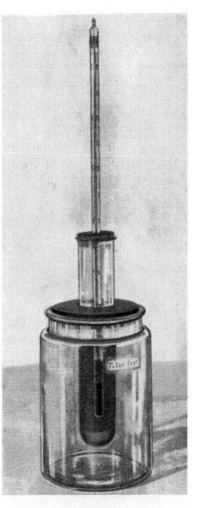

Fig. 21

8. The fatty acids are placed in the desiccator to cool and dry for 12 hours or more.

---

[1] If not quite clear the solution should be boiled for at least 30 minutes to ensure that the fat is completely saponified.

9. Melt the solidified fatty acids over a water-bath, and pour sufficient into the test-tube to half fill it. Insert thermometer carefully so that it is centrally placed in the liquid, and replace test-tube in the glass bottle.

10. Observe carefully when turbidity commences, or solid starts to settle out; stir the mass with the thermometer (*using great care as the bulb is very thin and delicate*) in one direction three times, and then in the contrary direction three times, then continuously so as to intimately mix the separating solid with the liquid, *but do not touch the sides of the test-tube*. The mercury of the thermometer is then carefully watched. It will first be found to fall regularly, or remain stationary, then a *sharp rise* will occur, after which it will remain stationary and then commence to fall again. The **top point of the rise** is taken as the **S.P. of the fatty acids**.

11. Repeat with more of the same fatty acids till two concordant results are obtained; determinations should not differ by more than $\frac{1}{10}°$ C.

Students should carefully wash out tube and cleanse thermometer with hot soft soap solution. Use the greatest care with the thermometer as these are of an expensive type and easily broken. Leave fatty acids in basin on bench. Cleanse out beaker, flask, and funnel as above.

REFRACTIVE POWER (4)

§ 1. Definition

The REFRACTIVE INDEX is a number expressing the ratio between the *velocity of light in air* and its velocity in some other medium. More particularly it expresses the ratio between the sine of the angle of *incidence* of a ray of light on the surface separating two media and the sine of the

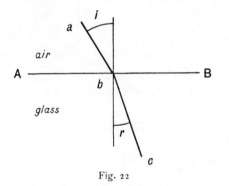

Fig. 22

angle of its *refraction*. This ratio is a *constant* for any two media. Thus, in the diagram, the line *AB* is the boundary between air and glass, and *ab* is a ray of light passing from air into glass where it is refracted along the line *bc*. Then whatever may be the value of the angle *i* the angle *r* will always be so related to it that

$$\frac{\sin i}{\sin r} = \text{a constant } n = \text{INDEX OF REFRACTION.}$$

Note.—If $i = 90°$, $\sin i = \sin 90° = 1$; $\therefore \frac{1}{\sin r} = n$, i.e. $\sin r = \frac{1}{n}$.

§ 2. Remarks

The index of refraction is thus a quantity which is a **constant** for a pure substance under standard conditions of temperature and pressure. The various glycerides differ in their refractive power[1], and its determination is thus a useful figure for ascertaining the purity of an oil or fat.

Many instruments have been devised by means of which this constant can be determined in a few minutes. Of these the more important for oils and fats analysis are the BUTYRO-REFRACTOMETER and the PULFRICH

---

[1] See Vol. I, p. 72.

refractometer, the latter instrument giving the refractive power with a high degree of accuracy, as well as the *dispersion* of the oil or fat if required.

In addition to these, there is the ABBÉ refractometer which may be used in place of the Pulfrich if desired, and which has the advantage of requiring only a drop of the liquid for the examination. It is useful when the oil gives figures outside the range of the butyro-refractometer scale, as for the analysis of *tung* and *rosin* oils.

The Pulfrich is a universal instrument, and by the use of several prisms of different refractivities can be used for the examination of any liquid.

The butyro-refractometer is now generally employed in analytical work for oils and fats owing to its simplicity and convenience.

All these instruments, formerly made by a German optical firm of great repute[1], are now manufactured in this country by Messrs Adam Hilger, of London.

The *oleo-refractometer* of *Amagat* and *Jean*, giving differential readings on an arbitrary scale, and employing a "standard oil" (sheep's foot oil), offers no advantages over the butyro-refractometer, and the results obtained are not susceptible of calculation to refractive indices. Further, it requires several c.c. of the oil for the examination.

[1] Carl Zeiss, Jena.

(a) THE PULFRICH REFRACTOMETER

§ 3. Remarks

In this instrument a ray of light enters the liquid (or solid) under examination parallel to the surface of a glass prism of refracting angle 90°. The ray then travels along the path indicated in the diagram.

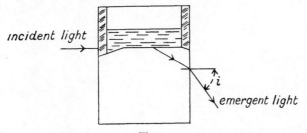

Fig. 23

By means of a reading telescope the angle i of emergence is measured and it can be shown that if $n = $ *index of refraction of the substance* under examination, and $N = $ *index of refraction of the glass prism*, then $n = \sqrt{N^2 - \sin^2 i}$; thence n can be calculated, knowing N and measuring i.

§ 4. Description

Fig. 24 represents the Zeiss[1] model of the **Pulfrich refractometer** together with apparatus for keeping the *prism* and oil or fat under examination at any desired temperature. Figs. 25 and 26 give a front view and a back view respectively of the refractometer, in which **c** is the *eyepiece of the telescope* through which the observation is made, **d** a *milled screwhead* for clamping the *divided circle* **g** so that it can be moved through a small angle by the *screw* **b**. **e** is a small *magnifier* for observing the coincidence of the *divided scale* with the *fixed vernier* **f**. **h** is a movable *right-angle prism* for refracting the light from a *sodium bunsen flame* through the oil or fat under examination contained in a *small cell* **i** resting on the *refracting prism* **j**. **k** is a Geissler *vacuum-tube* containing *hydrogen* for examining the oil, etc., with light of *C and F lines of spectrum*, which is focussed by the *lens* **l**, the height of which is adjusted by the *screw* **m**; the *shutter* **n** allows more or less light to pass from the vacuum-tube to the prism.

§ 5. Preparatory Arrangements[2]

(*a*) The **water pressure regulator** is connected with the water supply, and the former is joined up with the heating spiral, and

---

[1] The English instrument corresponds in its main features with this description.
[2] Students will probably find these matters attended to and can then omit this paragraph.

tubes from this are connected with the instrument as shown in the photograph. The heat is then regulated until a constant temperature of **40° C.** is obtained, which is the generally agreed standard for oils and fats determinations[1].

Fig. 24.   Pulfrich Refractometer: general view

(b) A MONOCHROMATIC LIGHT is obtained by placing salt on the burner supplied, and lighting the bunsen flame. It should give an even area of bright yellow illumination. It is then placed in the required position. This gives the refractive index for the sodium D line of the spectrum.

(c) If the C and F lines of the hydrogen[2] spectrum are to be

[1] See Vol. I, p. 72.
[2] Lithium, cadmium and mercury also give well-defined readings.

employed, the vacuum-tube is fixed in position and connected with the induction coil and battery.

Fig. 25. Pulfrich Refractometer: front view

§ 6. **Procedure**

1. DETERMINE THE ZERO ERROR OF THE INSTRUMENT.
   This is done in the following manner:
   Set the zero of the *vernier* to the zero of the *graduated circle*, place a light opposite the small window **a** in the *reading telescope*, and observe (by looking through the latter) the image of the *cross-wires* reflected from the face of the *refracting prism*. If the zero

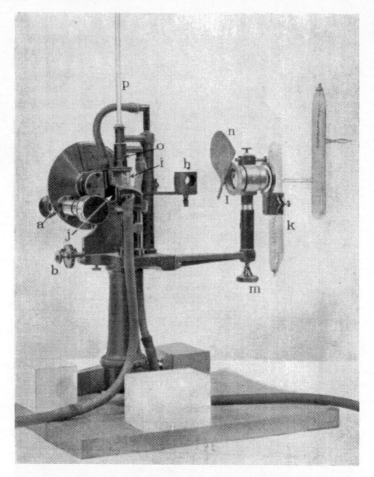

Fig. 26.   Pulfrich Refractometer : back view

Fig. 27

is correct the **image will correspond**[1] **with the cross-wires.** As a rule this is not the case.   If they do not correspond, turn the *screw* **b** so as to make the *image* of one of the wires coincide with its cross-wire.   Then take the reading of the vernier[2].   Repeat this with the other wire. The *mean* of these two readings must be added or subtracted from all subsequent observations in order to give the **corrected reading**.

[1] That is the wires themselves will exactly hide their reflected images.

[2] If students find any difficulty in reading the vernier, the demonstrator's assistance should be asked.

2. See that the oil is *clear*. If not, filter through dry, thick filter paper. If a fat or wax, melt at a gentle heat in a porcelain basin.
3. Place enough of the sample in the *cell* **i** to cover the surface of the refracting prism to a depth of 2—3 mm.
4. By means of the *milled head* **q** carefully lower the *heating tube* **o** (carrying the thermometer **p**) into the oil, and allow to remain for a few minutes until the sample has acquired the necessary standard temperature.
5. Loosen *screw* **d**, and, with the eye to *telescope* **c**, move *circle* **g** until the yellow band of light comes into the field of view.
6. Now clamp **d**, and by means of *screw* **b** bring the top sharp edge of the band of light to the *intersection* of the cross-wires.
7. Take reading of scale, using the magnifier to read vernier.
8. Displace the band of light, and bring again to the intersection of the cross-wires and observe reading again. Repeat, displacing the band in the *opposite direction*.
9. The *mean* of these readings is taken as final.
10. Add or subtract any error ascertained in (1) above and refer to tables supplied to ascertain the **refractive index.** The second column of the tables (n_D) signifies the D line of sodium light. The reading (in degrees) is given in the first column (i). The calculation is according to the previously mentioned formula :

$$n_D = \sqrt{N^2 - \sin^2 i}.$$

§ 7. **Example**

Suppose that the mean of the readings was 42° 15′. The part of the table concerned reads :

PRISMS I*b* AND IV*b*.

i	n_D	Δ_n
42° 0′	1·47751	
42° 10′	1·47653	9·8
42° 20′	1·47555	9·8

From the table n_D for 42° 10′ = 1·47653.

Δ_n, i.e. the difference of index for 1′ = 9·8 (in fifth decimal place) ;

∴ Δ_n ,, ,, ,, ,, ,, 5′ = 5 × 9·8 = 49.

Hence n_D for 42° 15′ = 1·47653 − ·00049
          = **1·47604.**

* The cell **i** need only rest on the prism and need not be cemented thereto ; it will be found that if adjusted properly to the surface of the prism it will retain the sample of oil without leakage for a long period ; moreover when the sample is finished with, the oil can be removed by inserting into the cell a piece of cotton wool which absorbs the oil, removing cell and cotton wool together and then carefully wiping prism with a piece of cotton wool moistened with ether. The prism will then be ready for a fresh sample. The cell and hot water tube can be similarly cleaned with cotton wool moistened with ether.

Since the scale readings of the butyro-refractometer are so largely used in oil analysis, the following table has been calculated giving the equivalent of scale readings for refractive indices within the range of oils and fats:

Table for Conversion of Refractive Indices n_D into Scale Readings of Butyro-refractometer.

Refractive Index n_D	Scale Reading	Refractive Index n_D	Scale Reading	Refractive Index n_D	Scale Reading
1·4450	29·8	1·4565	45·8	1·4680	63·2
1·4455	30·4	1·4570	46·5	1·4685	63·9
1·4460	31·0	1·4575	47·2	1·4690	64·7
1·4465	31·6	1·4580	48·0	1·4695	65·5
1·4470	32·3	1·4585	48·8	1·4700	66·3
1·4475	33·0	1·4590	49·5	1·4705	67·1
1·4480	33·8	1·4595	50·3	1·4710	67·9
1·4485	34·5	1·4600	51·0	1·4715	68·7
1·4490	35·2	1·4605	51·8	1·4720	69·5
1·4495	35·9	1·4610	52·5	1·4725	70·3
1·4500	36·7	1·4615	53·2	1·4730	71·1
1·4505	37·4	1·4620	54·0	1·4735	71·9
1·4510	38·0	1·4625	54·7	1·4740	72·7
1·4515	38·7	1·4630	55·5	1·4745	73·5
1·4520	39·5	1·4635	56·2	1·4750	74·3
1·4525	40·2	1·4640	57·0	1·4755	75·1
1·4530	40·9	1·4645	57·8	1·4760	76·0
1·4535	41·6	1·4650	58·5	1·4765	76·8
1·4540	42·3	1·4655	59·3	1·4770	77·6
1·4545	43·0	1·4660	60·1	1·4775	78·4
1·4550	43·7	1·4665	60·8	1·4780	79·1
1·4555	44·4	1·4670	61·6	1·4785	80·8
1·4560	45·1	1·4675	62·4	1·4790	81·5

The correction for temperature varies with each oil, fat or wax, and differs for the same number of degrees at different temperatures. The average correction is ·00038 for each degree C., the refractive index becoming less as the temperature rises.

(b) **THE BUTYRO-REFRACTOMETER**

§ 8. This instrument has become the standard for oils and fats determinations, owing partly to the ease and rapidity of its adjustments, and to the fact that only a few drops of the sample are required for the determination, but chiefly to the convenience of the scale readings as whole-number figures of comparison.

§ 9. Description

Reference to the figure (28) will show the construction of the instrument. It consists in principle of a *double prism*, between the surfaces of

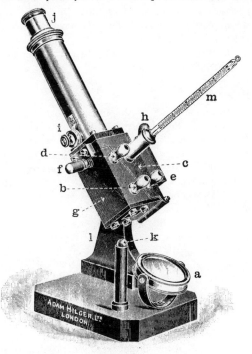

Fig. 28

which a fluid can be placed, and which is capable of being maintained at a uniform temperature. This is attained by passing a current of uniformly heated water through the instrument.

The light is projected into the instrument by means of the *mirror* **a**, and penetrates through the film of oil between the prisms. The border line of the total reflection is examined by means of a *fixed telescope* **j**, the objective of which is inserted in an adjustable slide, movable by the *micrometer screw* **i**. The reading of the instrument is taken on an *engraved scale*, which is placed in the focal plane of the objective, and which is focussed clearly by means of the upper lens of the eyepiece at **j**.

When daylight is used, the *border line* seen through the telescope is *colourless* in the case of pure butterfat (since the prism is achromatised for this reading), while at other parts of the scale it has a *coloured fringe*.

The use of a monochromatic light, such as the *sodium bunsen flame*, gives a clearly defined line, and is preferred for general work.

§ 10. Preparatory Arrangements

Adjust the constant temperature device as described in § 5 *a*. The water should enter at **b** and leave the instrument at **h**, **d** and **e** being connected by a piece of rubber tubing. A uniform temperature of **40° C.** is necessary unless the fat or wax is solid at this temperature. In such cases the determination is made about 2° above the melting point, and the result calculated[1] to 40° C.

Observe the thermometer constantly to make sure that the correct temperature is being maintained.

§ 11. Procedure

1. Grasp the instrument by the telescope carrier **l** (never by the telescope itself) with the left hand, open the prism by turning the *milled head* **f** half a turn, and allow the hinged portion to rest on **k**.

2. If the prism surfaces are not scrupulously clean, they should be very carefully wiped with a *soft linen handkerchief* (not cotton) dipped in a mixture of alcohol and ether. The surrounding metal parts should also be quite clean.

3. If the sample to be tested is clear, pour a few drops on to the prism surface on **c**, meanwhile holding refractometer so that this is horizontal. Taking now the hinged portion **g** close this over **c** and secure by turning the milled head **f** a half turn. If the oil is not quite clear it must be filtered through a dry, thick filter paper. If a fat or wax this should be melted in a test-tube at a *gentle* heat.

4. Place the instrument on the bench and arrange the sodium lamp so that the light is reflected by the *adjustable mirror* **a** through the instrument, and adjust the eyepiece lens so that the scale can be distinctly seen.

5. Allow a minute for the film of fat to attain the correct temperature, and then examine. If the line is not quite sharp and distinct it is probably due to air-bubbles or spaces, and in this case a fresh application of oil is made to the prism. It is often due however to the oil not having attained a uniform temperature, and a little further time must be allowed before examination.

6. Read off the scale figure for the line in whole numbers. Ascertain the first decimal place by the aid of the *micrometer screw* **i** in the following manner:

---

[1] The correction for butterfat is 0·55 scale divisions for each degree C.—that is, for each degree rise in the observation over 40·0°, 0·55 scale divisions must be added to the reading.

Note the position of the micrometer drum, and the reading it gives (0·21 in fig. 29). Then, with the eye to the telescope, rotate the milled head (thus moving the position of the border line) until the line is *just coincident* with the next unit division of the scale. Read the new figure on the drum and subtract from the first reading. This will give the decimal to be added to the whole number.

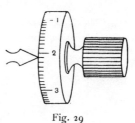

Fig. 29

7. Repeat all operations, using the *standard fluid* supplied with the instrument, and make any correction which is found necessary.

8. To convert scale readings into refractive indices the following formula has been calculated by *Roberts*[1]:

$$[n]_D = 1\cdot4220 + 0\cdot00142\,x\left(\cdot5753 - \frac{x}{1000}\right),$$

where x = scale reading.

Students should leave the prisms perfectly clean, using for the purpose a soft linen handkerchief and a little alcohol-ether mixture.

The following table, calculated by the authors, gives the refractive indices $[n]_D$ for corresponding scale readings.

Table for Conversion of Scale Readings (*Butyro-refractometer*) into Refractive Indices n_D.

Scale Reading	Refractive Index n_D	Scale Reading	Refractive Index n_D	Scale Reading	Refractive Index n_D	Scale Reading	Refractive Index n_D	Scale Reading	Refractive Index n_D
30·0	1·4452	40·0	1·4524	50·0	1·4593	60·0	1·4659	70·0	1·4723
30·5	1·4456	40·5	1·4528	50·5	1·4597	60·5	1·4663	70·5	1·4726
31·0	1·4460	41·0	1·4531	51·0	1·4600	61·0	1·4666	71·0	1·4729
31·5	1·4463	41·5	1·4535	51·5	1·4604	61·5	1·4669	71·5	1·4732
32·0	1·4467	42·0	1·4538	52·0	1·4607	62·0	1·4672	72·0	1·4736
32·5	1·4470	42·5	1·4542	52·5	1·4610	62·5	1·4675	72·5	1·4739
33·0	1·4474	43·0	1·4545	53·0	1·4613	63·0	1·4679	73·0	1·4742
33·5	1·4477	43·5	1·4549	53·5	1·4616	63·5	1·4682	73·5	1·4745
34·0	1·4481	44·0	1·4552	54·0	1·4620	64·0	1·4685	74·0	1·4748
34·5	1·4485	44·5	1·4556	54·5	1·4623	64·5	1·4688	74·5	1·4751
35·0	1·4488	45·0	1·4559	55·0	1·4626	65·0	1·4691	75·0	1·4754
35·5	1·4492	45·5	1·4563	55·5	1·4629	65·5	1·4694	75·5	1·4757
36·0	1·4495	46·0	1·4566	56·0	1·4633	66·0	1·4698	76·0	1·4760
36·5	1·4499	46·5	1·4569	56·5	1·4637	66·5	1·4701	76·5	1·4763
37·0	1·4502	47·0	1·4573	57·0	1·4640	67·0	1·4704	77·0	1·4766
37·5	1·4506	47·5	1·4577	57·5	1·4643	67·5	1·4707	77·5	1·4769
38·0	1·4510	48·0	1·4580	58·0	1·4646	68·0	1·4710	78·0	1·4772
38·5	1·4514	48·5	1·4584	58·5	1·4649	68·5	1·4713	78·5	1·4775
39·0	1·4517	49·0	1·4587	59·0	1·4653	69·0	1·4717	79·0	1·4778
39·5	1·4521	49·5	1·4590	59·5	1·4656	69·5	1·4720	79·5	1·4781

---

[1] *Analyst*, 1916, **41**, 376.

(c) THE ABBÉ REFRACTOMETER

§ 12. By means of this instrument direct measurements of the refractive index may be made of all liquids ranging between 1·3 and 1·7. An accuracy of 1 unit in the fourth decimal place is attainable. It is suitable for general mineral and fatty oil analysis, and for the examination of turpentine and essential oils.

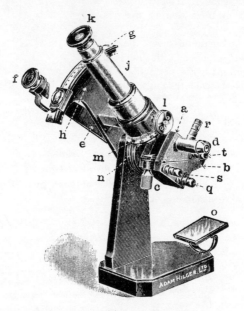

Fig. 30

§ 13. Description

a and **b** are *prisms* of dense flint glass, mounted in *water jackets*, which can be clamped together by turning the *milled head* **c**. The face of the upper prism in contact with the liquid is polished, while the corresponding surface of the lower prism is left grey to prevent the formation of false images. The *thermometer* screws into position close to the *nozzle* at **d** and records the temperature of the circulating water. The prism system can be rotated about a horizontal axis by means of the *arm* **e**, to which is attached the *reading lens* **f**, and the index which moves over the *scale* **g**. The *telescope* **j** is rigidly attached to the *sector* **h** carrying the *scale* **g**, and the whole system of sector and telescope together with the prisms can be turned about a horizontal axis, the extreme positions being determined by stops. The figure shows the instrument in the normal position for measuring liquids, light being reflected into the prisms by means of the *mirror* **o**. The *Abbé compensator* is attached to the end of the telescope, the prisms being rotated by the *milled head* **l** and their position recorded by the *scale* **m**. The lower prism can be rotated relatively to the scale by the *black milled ring* **n**.

§ 14. Preparatory Arrangements

In order to obtain correct readings, the instrument is checked by means of the glass test-piece supplied, having the refractive index engraved upon it. The reading is made as follows : The sector **h** is moved over against the forward stop, the prism jackets opened, and the instrument turned round so that the telescope points towards the light source. The test-piece is now placed with its larger polished face on the prism face, a drop of liquid of higher refractive index (mono-bromnaphthalene is recommended) being used to make the optical contact. The dividing line of the field is now achromatised by means of the compensator. The line is then set on the intersection of the cross-webs by moving the index arm, and the reading noted. If the latter does not agree with the value engraved on the test-piece, the prism jackets must be moved relative to the prism arm, three screws being gently loosened for this purpose and retightened when the setting is correct.

The instrument is connected with the thermostat, and a stream of water passed through the prisms at the required temperature. The water stream should enter at the nozzle **q** and leave at **r**, the other two nozzles, **s** and **t**, being connected by rubber tubing.

§ 15. Procedure

1. See that the thermometer has remained steady at the required temperature for 5—10 minutes.
2. Push sector **h** over to front stop, opening prism jackets, and moving index arm **e** if necessary until the face of the prism is horizontal.
3. Wipe the surface with a clean soft cloth, place a drop or two of liquid on, close the jacket over and fix by turning milled head **c**. Restore sector to normal position, and direct a beam of light through the telescope by adjustment of the mirror.
4. Focus the cross-webs sharply by means of *eyepiece* **k**. On turning the *index arm* **e** the dividing line of the field will, in general, be indicated by a patch of bright colour between the light and dark parts of the field. Remove this by turning the *milled head* **l**, when the dark half of the field will appear bounded by a sharp edge accompanied by a purplish border.
5. By a further movement of the index arm bring the sharp edge directly under the intersection of the cross-webs.
6. Read the refractive index directly on the scale, using the magnifier **f** for this purpose. The third decimal place is engraved, and the fourth decimal is judged by inspection.

    To obtain the corresponding *scale readings* of the butyro-refractometer refer to table on page 52.

§ 16. Correction for Temperature

The average correction for temperature for oils, fats, etc. is ·00038 for each ° C. rise. If the observation therefore is taken at a temperature $x°$ over 40·0° C., the correction necessary is $+ (x \times ·00038)$. (For *turpentine* the factor is nearer ·00035.)

Students must leave the prism surfaces perfectly clean, using for this purpose a soft linen handkerchief and a little alcohol-ether mixture. The prisms should not be rubbed but only touched lightly.

THE DISPERSIVE POWER

§ 1. Definition

The difference between the refractive indices of any two rays of the spectrum of any refracting medium is termed the *dispersion* of that medium, while the DISPERSIVE POWER (ω) of a substance is expressed by the ratio of the coefficient of dispersion to the index of refraction of the mean ray minus unity. Thus if n_C, n_D and n_F represent the refractive indices of a substance for the C, D and F lines of the spectrum, then

The *Dispersion* of the substance $= n_F - n_C$ for the F and C lines.

The *Dispersive Power* $(\omega) = \dfrac{n_F - n_C}{n_D - 1}$.

§ 2. Remarks

The authors have examined the dispersion of most of the fatty oils, and the results are given in diagrammatic form, and also in the table following. They found the figure of much less analytical value than the refractive index. The drying and semi-drying oils are broadly differentiated from the non-drying oils and fats, but the separation is not very marked. Coconut and linseed oils are however exceptional, the former being much lower and the latter higher than the average. It is thus evident that the presence of glycerides of acids of low molecular weight decreases the dispersion, whilst unsaturated glycerides increase the figure. A remarkable exception to the general rule is Tung (Chinese wood) oil, which shows a very high dispersion indeed, this being perhaps its most characteristic property.

The authors find that FREE ACIDITY has little effect on the dispersion, the figure being slightly raised (about ·00001 for each five per cent. free acidity). OXIDATION of an oil increases the dispersion, while heating ("POLYMERISATION") produces a marked decrease. This affords a means of distinguishing "oxidised" from "polymerised" oils. "ELAÏDINS" give much lower dispersions than the oils from which they are produced.

The effect of the temperature of the oil during the observation is not very marked. Rise of temperature decreases the figure, apparently about ·00002 for each degree C. rise. [For further details the original paper should be consulted, *Analyst*, 1918, September.]

§ 3. Apparatus

The Pulfrich Refractometer is the most generally suitable for the dispersion, although readings may also be obtained on the Abbé instrument. The source of light most generally employed is that obtained by an electric discharge of high voltage through a tube of hydrogen under low pressure (about 2 mm.). This gives red and blue rays corresponding with the C and F lines of the spectrum.   A mercury or cadmium vapour lamp may also be employed.

§ 4. **Procedure**

1. Fit up the instrument exactly as for the determination of the refractive index (page 49), using the hydrogen tube, connected with a suitable induction coil, and also the sodium lamp.
2. Adjust constant temperature device to 40·0° C., place the oil in the cup on prism, lower the heating chamber into the oil, and obtain reading for the D line as directed on page 51.
3. Swing aside the reflector, and focus the light from the hydrogen tube on to the prism through lens **l** by means of screw **m**. Obtain reading of the red line of light—the top margin of the ray is taken, and should be perfectly sharp and distinct.
4. Turn the screw **b**—fixed to the small graduated drum—till the blue line comes into view, and bring it to the intersection of the cross-wires as before.   Then read from the graduated drum the angle through which the drum has been turned to bring the cross-wires

Dispersions of oils at 40·0° *C.* [*Fryer and Weston*].

Oil		n_D	$n_F - n_C$	$\omega = \dfrac{n_F - n_C}{n_D - 1}$	$\dfrac{1}{\omega}$
Menhaden	...	1·47361	·00979	·0207	48·4
Shark liver	...	1·46849	·00955	·0204	49·0
Seal	...	1·47018	·00962	·0205	48·9
Whale ...	...	1·46630	·00918	·0197	50·8
Perilla ...	...	1·47527	·00984	·0207	48·3
Linseed...	...	1·47379	·01032	·0218	45·8
Tung ...	...	1·51256	·01904	·0371	26·9
Hemp ...	...	1·47404	·00980	·0206	48·4
Walnut ...	...	1·47054	·00985	·0209	47·8
Poppy ...	...	1·46984	·00978	·0208	48·0
Niger ...	...	1·46968	·00935	·0199	50·2
Sunflower	...	1·47211	·00973	·0206	48·5
Maize ...	...	1·46711	·00938	·0201	49·8
Cotton ...	...	1·46535	·00910	·0195	51·1
Sesamé ...	...	1·46650	·00908	·0194	51·3
Rape ...	...	1·46770	·00936	·0200	50·0
Peach kernel	...	1·46439	·00910	·0196	51·0
Almond...	...	1·46403	·00890	·0192	52·1
Arachis...	...	1·46431	·00878	·0189	52·9
Olive ...	...	1·46184	·00862	·0187	53·6
Castor ...	...	1·47194	·00897	·0190	52·7
Cacao-butter	...	1·45724	·00853	·0186	53·6
Palm kernel	...	1·45034	·00812	·0180	55·5
Coconut	...	1·44924	·00751	·0167	59·8
Lard ...	...	1·45928	·00851	·0189	53·8
Butterfat	...	1·45427	·00830	·0182	54·7
Sperm ...	...	1·45814	·00864	·0188	53·0

Diagram of dispersive power of oils and fats.

$$\left(\frac{n_F - n_C}{n_D - 1}\right) 40\cdot0°\ C.$$

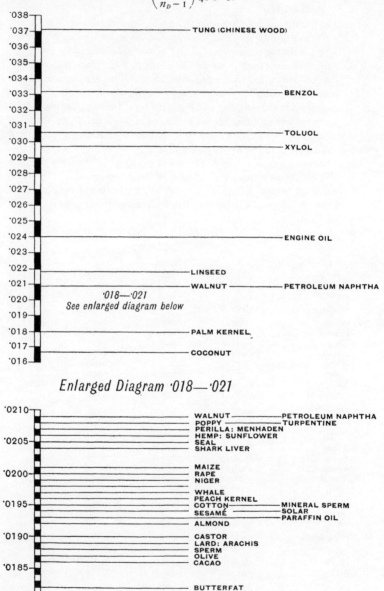

·038 —
·037 — TUNG (CHINESE WOOD)
·036 —
·035 —
·034 —
·033 — BENZOL
·032 —
·031 — TOLUOL
·030 — XYLOL
·029 —
·028 —
·027 —
·026 —
·025 —
·024 — ENGINE OIL
·023 —
·022 — LINSEED
·021 — WALNUT ——— PETROLEUM NAPHTHA
·020 — *018—·021*
·019 — *See enlarged diagram below*
·018 — PALM KERNEL
·017 — COCONUT
·016 —

Enlarged Diagram ·018—·021

·0210 — WALNUT ——— PETROLEUM NAPHTHA
        POPPY ——— TURPENTINE
        PERILLA: MENHADEN
        HEMP: SUNFLOWER
·0205 — SEAL
        SHARK LIVER

·0200 — MAIZE
        RAPE
        NIGER

        WHALE
        PEACH KERNEL
·0195 — COTTON ——— MINERAL SPERM
        SESAMÉ ——— SOLAR
                ——— PARAFFIN OIL
        ALMOND

·0190 — CASTOR
        LARD: ARACHIS
        SPERM
        OLIVE
·0185 — CACAO

        BUTTERFAT
·0180 — PALM KERNEL

from the red to the blue line. It will be noted that the drum is divided into 200 parts, and that one complete rotation of the drum moves the telescope through $\frac{1}{3}$ degree, or 20', therefore 1 division of the drum rotates it through 0·1'.

It frequently happens that no blue line is visible, being absorbed by the oil. In this case the oil must be bleached, preferably by exposure to light.

§ 5. **Calculation**

The refractive indices are obtained by reference to the tables supplied with the instrument, subtracting the correction for the C line and adding the correction for the F line. A temperature correction may also be necessary according to the temperature at which the tables have been calculated.

The dispersive power (ω) is then ascertained, according to the formula

$$\frac{n_F - n_C}{n_D - 1}.$$

The reciprocal of this figure is frequently employed in dispersive measurements (being a whole number).

§ 6. **Example**

Suppose readings of an oil of $n_D = 1·48387$ were as follows:

angle i for red line $C = 40° 30'$,

and the drum rotated through 6·3' to bring blue line to intersection of cross-wires, therefore

angle i for blue line $F = 40° 30' + 6·3' = 40° 36·3'$.

The index of refraction is now obtained from the tables as follows:

i	n_D	Δ_n	Correction for		
			C	F	G
40° 0'	1·48917		536	1348	2498
40° 10'	1·48820	9·7	536	1349	2500
40° 20'	1·48724	9·6	536	1350	2501
40° 30'	1·48627	9·7	537	1351	2503
40° 40'	1·48531	9·6	537	1352	2504

$$n_C = 1·48627 - ·00537 = 1·48090,$$
$$n_F = [1·48627 - (9·6 \times 6·3)] + ·01351$$
$$= [1·48627 - ·000604] + ·01351 = 1·49908.$$

Hence the dispersion $F - C$

$$= 1·49908 - 1·48090 = ·01818,$$

and dispersive power (ω) $= \dfrac{·01818}{1·48387 - 1} = \mathbf{·0376},$

and reciprocal $= \dfrac{1}{·0376} = \mathbf{26·8.}$

VISCOSITY (5)

§ 1. Definition

Viscosity may be defined as the resistance which the particles of a liquid offer to their free motion one over the other.

If a vessel of water is tilted, the surface of the water remains level, and on allowing the vessel to fall back to its original position, the water oscillates from side to side. After a short lapse of time however this motion ceases. The cessation is due principally to the viscosity of the water. In the case of highly viscous liquids, such as glycerin or castor oil, no oscillatory motion occurs at all, the liquid simply resuming slowly its original position.

§ 2. Remarks

All liquids exhibit viscosity to a greater or less degree, and this quality is independent of the density of the liquid; e.g. *glycerin* has a S.G. of 1·26, yet it is vastly more viscous than *mercury* with a S.G. of over 13½.

The coefficient of viscosity is the force required to move in opposite directions two surfaces of a liquid 1 sq. cm. in area, and 1 cm. distant with a velocity of 1 cm. per second. This is termed the **absolute viscosity** (represented by the Greek sign η), and it has been measured by determining the rate of flow of liquids through capillary tubes. It is calculated from *Poiseuille's* formula (see Vol. I, page 74).

As the determination of the absolute viscosity in this manner requires considerable skill and time for its performance it has been customary to ascertain the time taken by liquids to flow through a standardised orifice of relatively large dimensions.

There are however some grave objections to this practice. In the first place the rate of outflow is influenced by the **density** (S.G.) of the liquid, as well as by its viscosity. Again, when correction has been made for the density, the efflux period is proportional to the viscosity only in the case of liquids of *relatively high viscosity*. Further, as there are at least four "standard" instruments for efflux determinations, all of differing construction, the figures obtained are not even comparable, except within the margins stated. Another unfortunate complication which has been introduced is the multiplying of varying degrees of temperature at which determinations are made. It appears therefore very desirable that in future only data which are **strictly comparable** should be allowed in technical work, and these to be obtained at an agreed standard temperature.

Archbutt and Deeley have shown that it is a simple matter to standardise any type of efflux viscometer so that it may be used to obtain the true viscosity in **absolute terms**. They have ascertained the absolute viscosity of solutions of glycerin of varying densities, and it is only necessary to construct a chart for each instrument showing the rate of efflux compared with these figures for similar solutions. The chart once complete, the determination of the viscosity in absolute terms is merely a matter of simple calculation.

As stated in Volume I, page 74, the authors are of opinion that for technical work results expressed in whole numbers are preferable, being more easily memorised and compared. They propose therefore that for technical purposes the absolute viscosity figure be multiplied by 100, thus ($\eta \times$ 100).

The question of the **standard temperature** for the determination is a very important one. In the first place the temperature chosen should be such that it can be employed for all oils within reasonable limits of time for the determination.

Oils such as castor and most of the "cylinder" oils are too viscous at ordinary room temperatures for testing. It is also desirable that regard should be paid to the **actual working conditions** under which the oil is used, so that the figure shall have some practical significance. It has been shown[1] that even when the heating of bearings is very small and the coefficient of friction low, the temperature of the oil film is upwards of 8° C. above that of the surrounding parts, while it is probable that much higher temperatures than this obtain in most cases. Viscosity determinations at lower temperatures than this would therefore appear to have little practical value.

Again, it has been usual in this country to use the Fahrenheit scale for the efflux period numbers, but it is fairly evident that a viscosity figure of true significance should fall into line with other physical data in the employment of the *Centigrade* scale for its determination.

After careful consideration of the foregoing, the authors suggest that **40° C**. is probably the most suitable temperature as a standard (as already employed in the determination of the refractive indices of oils).

There is a further important point to consider. In the case of oils which are employed for lubrication at HIGHER TEMPERATURES, as e.g. in steam cylinders, and the water-cooled cylinders of internal combustion engines, an indication of the viscosity under such working conditions is very desirable. There is also the expression of the well-known fact that although all oils become less viscous as the temperature rises, they do not all lose their viscosity to the same degree, and this is an important factor in actual practice. In addition therefore to the standard determination of viscosity, the authors propose that a further figure is ascertained, conveniently termed the **Viscosity Ratio Number**. This represents the ratio between the absolute viscosity at standard temperature (40° C.) and at a higher temperature, that of *boiling water* being plainly the most

[1] Osborne Reynolds.

suitable (100° C.), and is obtained by dividing the former figure with the latter. Thus

$$\textbf{Viscosity Ratio number} = \frac{\eta \text{ at } 40°}{\eta \text{ at } 100°} \textbf{ C.}$$

The viscosity at 100° C. is therefore immediately ascertainable by dividing the standard figure by the ratio number, while the latter would also express the proportional reduction in viscosity of an oil[1] at the higher temperature.

The following table gives figures for a few oils :

Table of Viscosities ($\eta \times 100$) at 40° C. and of Viscosity Ratio numbers $\left[\dfrac{\eta \text{ at } 40° \text{ C.}}{\eta \text{ at } 100° \text{ C.}} \right]$ **or** $\eta \dfrac{40°}{100°}$ **C.**

Liquid	Viscosity	Viscosity Ratio no.
Water	0·657	2·31
Sperm oil	17	3·7
Veg. Oils		
Olive	34	4·8
Rape	40	5·0
Castor	248	15
Mineral Oils		
Russian spindle	28	6·5
,, medium machinery	68	10·3
American spindle (1)	15	4·4
,, ,, (2)	21	5·4
,, light machinery	30	6·1
,, engine oil (1)	51	8·0
,, ,, (2)	64	9·1
,, axle	81	9·8
,, heavy machinery	115	10·2
,, cylinder oils (1)	216	11·6
,, ,, (2)	380	15
,, ,, (3)	580	20
,, ,, (4)	900	25
,, ,, (5)	1000	26

§ 3. Efflux instruments

There are four of these generally employed, viz.

(*a*) The Redwood viscometer.

(*b*) ,, Engler   ,,

(*c*) ,, Saybolt   ,,

(*d*) ,, Coleman-Archbutt viscometer.

A description of the REDWOOD instrument will serve for the first three, since they only differ in constructional details and not in the method of their employment. The last is a special type and will be described separately.

---

[1] Full information on this point can, of course, only be expressed by the construction of the curve obtained by plotting the results of viscosity determinations of an oil at varying temperatures.

§ 4. The Redwood Viscometer

The photograph shows the viscometer ready for use. The instrument

Fig. 31. Redwood Viscometer

consists of a silvered-copper cylinder at the bottom of which is an agate plug bored through to form the standard orifice through which the liquid flows, and which is closed by a metal ball. The cylinder is filled to the gauge mark and a thermometer **e** is suspended in the oil to a constant depth by the clip **f.** Paddles are fixed to a cylinder which slips over the oil vessel and is provided with a handle **h** for stirring and a tube **i** for holding a thermometer. The oil vessel is mounted in a copper jacket **j** supported by three legs which are provided with levelling screws. A spirit-level is also provided.

The EFFLUX TIMES are determined as follows :

Procedure

1. Set up the instrument and place spirit-level on the top: level by means of adjusting screws on legs. Examine jet and see that this is perfectly clear[1] and dry, and also that cylinder is dry and clean.

2. Fill the jacket with water to a point just above the gauge mark, and heat to necessary temperature, **40° C. (104° F.).**

3. Fill the vessel with the liquid to be tested to the point of the gauge.
   It saves time to heat the liquid previously to the temperature required (40° C. = 104° F.).

4. Place a graduated **50 c.c.** flask under the orifice and then rotate the paddles, and adjust the source of heat till the water stands at **40·5° C.**

5. When the liquid under test is exactly at **40° C.** raise the ball, and with the other hand start a stop-watch ; then hang ball on clip **f.** Keep a close watch on the temperature of the water jacket during the running out of the liquid. This should now be allowed to fall to **40° C.** and kept at exactly that point till the determination is finished. When the liquid has filled the flask to exactly the **50 c.c.** mark, stop the watch and close the orifice. Note the time taken for the outflow of the **50 c.c.** of liquid.

6. Repeat until two successive times are exactly concurrent.

Students should leave viscometer empty, but need not use soap to cleanse oil chamber, which may be wiped with a soft rag. The thermometers must be carefully handled and returned clean to the racks provided for them.

§ 5. Standardising

In order that the viscometer shall be available for obtaining **true viscosities** it must be standardised in the following manner[2]:

1. Prepare eight[3] solutions of purest glycerin in distilled water by weighing the glycerin carefully and thoroughly mixing together. The solutions are placed in stoppered bottles, 250 c.c. of each being a suitable quantity.

[1] Jet should be very carefully cleaned with a dry feather. It is obvious that the accuracy of the instrument is dependent on the jet.

[2] See Archbutt and Deeley, *Lubrication and Lubricants*, p. 177 *et seq.*

[3] Additional solutions of intermediate viscosity should also be made up if the viscometer is to be employed for specially accurate work so as to obtain more points on the curve. (Two other points are indicated on the curve shown.) Other ranges of solutions may be chosen for oils between narrower limits of viscosity.

per cent.

						per cent.
(i)	Glycerol 110 grms make up to 250 c.c. with water					(44·0)
(ii)	,, 142·5 ,,	,,	,,	,,		(56·9)
(iii)	,, 165 ,,	,,	,,	,,		(66·0)
(iv)	,, 177·4 ,,	,,	,,	,,		(71·0)
(v)	,, 188·2 ,,	,,	,,	,,		(75·3)
(vi)	,, 201 ,,	,,	,,	,,		(80·4)
(vii)	,, 227 ,,	,,	,,	,,		(90·8)
(viii)	,, 237 ,,	,,	,,	,,		(94·8)

2. Carefully ascertain the S.G. of each of these at $\frac{20°}{20°}$ C. by the bottle method. General details of the method are given on page 27, but the following special points must be noted:

  (a) Fill the bottle at any temperature below 18° C., and then place the bottle (without stopper) in a beaker of water kept at 20° C. until no further expansion takes place. The bottle must be almost completely immersed and care must be taken that no water splashes in. The stopper is then inserted and the bottle carefully dried and weighed to the nearest milligramme.

  (b) Since the bottle is standardised for water at 15·5° C., a blank should be performed with distilled water at 20° C., and the specific gravity calculated by dividing the weight of the glycerin solutions by net weight of the water.

3. Determine the efflux times for each solution at **20° C. (68° F.)** as described in § 4. Do not leave the stronger solutions open to the air except when making the actual tests, as they *rapidly absorb moisture* from the atmosphere.

4. Find also the efflux time of pure water at 20° C.

5. Find the viscosity of each solution by reference to the table on page 265, Vol. I, where the viscosities of solutions of glycerin of varying S.G. are given[1]. The viscosity of water is ·01028 at 20° C.

6. Ascertain now the density (d) of each solution by multiplying the S.G. $\frac{20°}{20°}$ C. by 0·998259 (density of water at 20° C.). The viscosity (η) and the efflux times in seconds (t) are already known.

  It only remains to determine the variation of the viscometer for liquids of different viscosities. This figure is K and is ascertained as follows:

$$K = \frac{\eta}{td},$$

i.e. the viscosity of the water and of each solution is divided by the product of the density of each into the time in seconds.

  Suppose that the efflux time of one of the solutions is **87·2 seconds**, its viscosity **0·2360** and the density **1·1827**. Then

$t \times d = 103·1$ and $\frac{\eta}{td} = ·00229 = K.$

[1] Due to Archbutt and Deeley.

7. Having now the values of K for each solution of known viscosity, a diagram is plotted showing the values of K for all intermediate values of $t \times d$ (i.e. efflux times × densities). Proceed in this manner:

(a) Take a large sheet of squared paper and mark off along a horizontal line the values ascertained for td for water and the first four solutions (up to about 250), along the vertical line mark off the values of K from o to ·004.

(b) Mark off a cross at the positions on the scale of the five determinations, and carefully join the points in an even curve.

(c) Take another sheet of squared paper, and in a similar manner mark off on a horizontal line the values o—4000td, and on a vertical line the same values for K as before. Mark off now all points for the nine determinations and carefully join in an even curve. *Small scale* diagrams are shown opposite of what is required:

§ 6. **Determination of Viscosity** $(\eta \times 100)$ 40° C.

As previously stated, this is the *absolute* figure multiplied by 100 to give a whole number. It will be strictly comparable with results obtained in any other make of viscometer standardised as in § 5. Archbutt and Deeley have shown that a special correction is necessary for oils below 40.

In this case it is more correct to take the value of K corresponding to $\frac{7}{8}td$ (see below).

Procedure

(a) *Oils with viscosity over* 40.

1. Determine efflux time (t) as described in § 4.

2. Ascertain density of oil (d) at 40° C. as in section 2 (a) of § 5. The S.G. at $\dfrac{40°}{15\cdot5°}$ C. is taken by bottle, and this figure multiplied by 0·99908.

3. Multiply t by $d = td$.

4. Find on scale the value of K corresponding to td.

5. Multiply td by 100K.
   The result = $(\eta \times 100)$ at 40° C.

(b) *Oils with viscosity below* 40.

1, 2 and 3 as before.

4. Find value of K on scale corresponding to $\frac{7}{8}td$.

5. Multiply td by 100K.
   = $(\eta \times 100)$ at 40° C.

Examples

(A) A sample of sperm oil gave figures:

1. Efflux time (t) 40° C. = 93 seconds.

2. The S.G. at $\dfrac{40°}{15\cdot5°}$ C. was 0·8692,

∴ density (d) = ·8692 × ·99908 = ·8684.

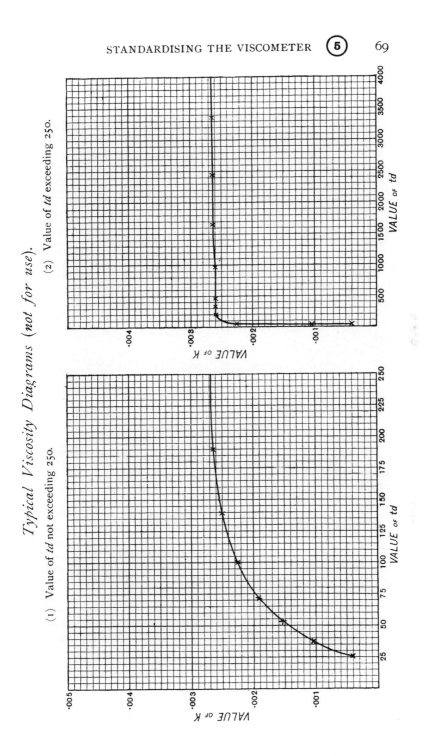

Typical Viscosity Diagrams (not for use).

(1) Value of *td* not exceeding 250.

(2) Value of *td* exceeding 250.

4. Take number of seconds required for the oil to reach zero mark, keeping jacket meanwhile uniformly at 40° C.

5. For determinations at 100° C. for ratio number, the jacket is first heated by pouring in hot water, and the funnel is then replaced by a delivery tube from a boiling-can in which steam is generated. The oil is heated to 100° C. and poured in, and the top of the tube protected with a glass cap.

Standardisation of the Viscometer

This is accomplished exactly as before stated (§ 5) for the Redwood instrument. Separate scales must be made for the three volumes 100, 50 and 25, and the K figures determined for each.

NON-EFFLUX VISCOMETERS

In the case of oils containing suspended matters, such as deflocculated graphite, and with highly viscous oils, as for example the heavy cylinder oils, it is not practicable to obtain the viscosity by efflux measurements. For these particular cases a non-efflux type of instrument is necessary. The best and most reliable of these is the DOOLITTLE TORSION VISCO-METER.

DOOLITTLE TORSION VISCOMETER

Description

Figs. 33 and 34 show the construction of the instrument. The principle concerned is the retardation of the swing of a torsion pendulum which is immersed in the liquid under examination. The more viscous the liquid, the greater is its slowing action on the pendulum. The specific gravity has no influence on the result. The apparatus consists of a steel cylinder 2 inches high by $1\frac{1}{2}$ inches diameter suspended by a steel wire from a fixed point, the wire passing through, and firmly attached to, a horizontal graduated disc. The cylinder is immersed in a vessel containing the oil and this vessel is surrounded by a jacket containing water or other liquid which can be heated to the necessary temperature.

Procedure

1. Fill cylinder with the oil to be examined.
2. Suspend the pendulum in the oil, and by means of a thumb-screw at top of instrument bring the zero on disc to the index mark, so that there is no torsion on the wire.
3. Fix the disc, and by means of the thumb-screw twist the wire through a complete circle (360°) and fix it also.
4. When the oil is at the temperature required, release the disc, and observe its angular oscillation, (1) to right, (2) to left, (3) to right again.
5. Again clamp disc and twist wire through 360° but this time in the opposite direction to that in 3.
6. Release disc and again take three observations of the angular oscillation.

Calculation

The pendulum swings from 360° to zero, and then continues to, say, (1) 340°, returning again to zero, and continuing in the other direction to, say, (2) 332°, returning to zero, and swinging back to, say, (3) 324°.

Neglecting the first arc described from 360° to 340°, we obtain :

1st complete swing = 340° + 332° = 672°,

2nd   „   „   = 332° + 324° = 656°.

Hence retardation =   16°.

Fig. 33.   Doolittle Torsion Viscometer

Calculate in the same manner the swing in the opposite direction (5 and 6 above), and take the mean of the two results.

Standardisation

Each instrument is standardised against solutions of pure cane sugar. A curve or table is supplied with the instrument showing the concentration of cane sugar solutions equivalent to all degrees of retardation.

Fig. 34. Doolittle Torsion Viscometer

Expression of Results

These are expressed in terms of the number of grammes of pure cane sugar, which, contained in 100 c.c. of the syrup at 60° F., will give at 80° F. the same retardation as the oil. The figures are obtained from the retardation in degrees on reference to the table supplied, or the instrument may be standardised by the student, employing varying percentages of cane sugar in pure water for the purpose.

Students must leave viscometers empty, but the oil chamber should not be washed out with soap solution, being simply wiped clean with a soft rag. The pendulum can be treated in the same manner.

SOLUBILITY[1] ⑥

§ 1. Definition
The solubility of oils in various solvents is a constant, depending on the nature of the glycerides composing the oil.

§ 2. Outline of Method
The method usually employed for measuring the differential solubility of oils in selected solvents is to note the temperature at which the solution, obtained by heating together equal volumes of the oil and solvent, shows, on cooling, the first signs of turbidity, thus indicating dissolution.

§ 3. Remarks
Various liquids differ in their solvent action on oils and fats. An exception to all other oils in its behaviour towards solvents is CASTOR OIL (see Vol. I, page 143). Solvents may be divided into three groups, according to their action on oils and fats.

GROUP 1.
  Solvents dissolving oils and fats in every proportion :
    Ethyl ether and other ethers.
    Carbon disulphide.
    Carbon tetrachloride.
    Chloroform and other chlorinated hydrocarbons.
    Petroleum ether and other hydrocarbons.

GROUP 2.
  Solvents only partially dissolving oils and fats (usually containing a
    hydroxyl group):
    Ethyl alcohol.
    Methyl alcohol.
    Acetic acid and some other organic acids.
    Acetone.
    Carbolic acid and other phenols.

GROUP 3.
  Solvents in which oils and fats are practically totally insoluble :
    Water.
    Glycerin, etc.

CASTOR OIL is soluble in the solvents of group 2, and insoluble in petroleum ether and hydrocarbons generally. Other oils vary to some extent in their solubility in solvents of the intermediate class. This is due to the *differing solubility* of the glycerides of which they are composed.

[1] See Fryer and Weston, *Analyst*, 1918, **43**, 4—20.

In the diagrams on pages 84, 85 it is clearly seen that

1. Oils containing glycerides of *low molecular weight* (e.g. coconut) are more soluble than the average.
2. Oils containing glycerides of *high molecular weight* (e.g. rape) are less soluble.
3. Oils containing a high proportion of *unsaturated* glycerides show increased solubility.

The authors have found[1] that there are two factors which, if not allowed for, entirely destroy the reliability of the estimation of solubility in these solvents. These are:

1. FREE FATTY ACIDS.
   These lower the turbidity temperature, increasing the solubility of the oils.

2. MOISTURE.
   This raises the turbidity temperature, decreasing the solubility.

In the case of oils and fats containing over 10—15 per cent. free fatty acids (in terms of oleic) these must be removed by shaking with 90 per cent. alcohol, separating, and drying the oil. In cases where the acidity is less than this, the figure may be calculated for the neutral oil by making the necessary allowance for the lowering of the turbidity temperature. The authors have estimated the fall in the temperature produced by one per cent. acidity in oils typical of the various classes. These figures may be used in calculating the results for neutral oil for any oils of these classes.

The question of variation of the figure due to moisture, and alterations in the concentration of the solvents used, is met by the use of a **standard oil** for the test, ALMOND OIL being chosen for this purpose on account of its purity (English expressed[2]), low acidity and freedom from stearine.

If the pure glyceride *olein* were easily prepared, this would probably be the ideal standard for the purpose. Almond oil consists almost entirely of olein, and if any difficulty exists in obtaining a pure specimen, the oil may be expressed or extracted in the laboratory.

It has been suggested to standardise the acetic acid by means of titration with alkali, or by the solidifying point. Neither of these methods is sufficiently delicate for the purpose, and since the acid is extremely hygroscopic it appears hopeless to attempt its standardisation.

We term the results corrected for these variations and adjusted to the standard oil the **True Valenta** (acetic acid) and **Turbidity Value** (alcohol reagent) (see below).

§ 4. Solvents

The solvent originally used by *Valenta* was ACETIC ACID, and the authors have found that this is the best to employ in most cases, owing to the greater degree of discrimination which it gives with most oils. In this case, the acidity of the oil or fat must be independently determined.

---

[1] *Analyst*, 1918, **43**, 4—20.
[2] Obtained from any druggist.

As acetic acid is very prone to change its value, owing to absorption of moisture, it is necessary to make an observation with the standard oil side by side with the oil under examination. The acetic acid is kept at or within 10° either way of **80° C.** test to neutral almond oil.

The authors found that an excellent solvent for rapid sorting purposes was a mixture of approximately 92 per cent. ETHYL ALCOHOL[1] and an equal volume of AMYL ALCOHOL. This is then standardised by addition of the requisite amount of water until it gives **70·0° C.** turbidity temperature with the standard oil. It is not liable to change its value, but should be checked from time to time. After the turbidity temperature has been ascertained, the *same sample* may be used for determining the acidity of the oil, and the necessary correction for the acidity at once made.

§ 5. Preliminary Operations

STANDARDISATION OF THE SOLVENT.

(A) ACETIC ACID.

The "glacial" acid is frozen, and the crystals drained at a temperature of 15° C., until no further liquor runs off. They are then melted, and if within 10° below or above 80° test with standard almond oil[2] (see below), the acid is suitable for use. If above 90° test, a further freezing and draining of the crystals is necessary. If below 70° test, water is added to the acid to bring it to the requisite degree of concentration.

The amount of water necessary is about **·029 per cent.** for each degree C. rise required.

(B) ALCOHOL REAGENT.

Mix equal volumes of 92 per cent. ethyl alcohol and pure amyl alcohol. Determine the turbidity figure with standard almond oil, and make the necessary addition of water to bring it up to 70·0° C.

The acid value (in terms of oleic) must be ascertained, and for every **one per cent. acidity** the standard temperature lowered **2·07°.** Thus, in the case of almond oil of acidity 1·5 per cent., the alcohol is adjusted to give a turbidity temperature of 67° C. (66·9).

The amount of water required to be added will be about **0·11 per cent.** for each degree C. rise necessary. When once adjusted the reagent will not tend to vary, though it should be checked from time to time.

(C) THE STANDARD OIL.

This should be "English expressed" almond oil, and should be tested for Iodine Value (below 100) and by *Bieber's* test (page 140) (negative). It should be kept in a well-corked bottle, and the acidity re-determined at intervals.

---

[1] The authors found that industrial (non-mineralised) methylated spirit was suitable for this.

[2] The *acidity* must be determined and allowed for as described later.

§ 6. Preparation of the Oil

MOISTURE is removed by filtration through a dry, thick filter paper, if the oil, on heating to 100° C. in a water-bath, is not *perfectly clear*. In any case, the oil, placed in a test-tube immersed in boiling water, is filtered through dry cotton wool. This is accomplished by fitting a wad of the wool closely into the tube, and pushing this slowly in a spiral direction through the oil. A stout piece of wire bent into the form of a plunger is used for this purpose. It should just fit the tube easily. (See fig. 35.)

§ 7. Procedure
(*a*) ACETIC ACID.

The acid is preferably kept in a vacuum flask (of the kind readily purchased at a low figure nowadays). It will be then very unlikely to freeze during the night in cold weather. If this is not available an ether

Fig. 35

bottle will suffice, but the acid must be melted when frozen[1], and will then probably change in value. A 2 c.c. pipette with a long stem is kept permanently immersed in the acid, being held tightly in a good cork, the open end kept closed by means of a piece of rubber tube provided with a cap of solid glass rod, the whole forming a close cover. It is thus unnecessary to suck the acetic acid into the pipette for measurement, and accession of moisture from this cause is avoided.

For the test, a TEST-TUBE is used of such a diameter that the thermometer bulb is *completely covered* by 4 c.c. of liquid.

The thermometer is inserted tightly into a cork fitting to the test-tube, a small groove being cut in the side of the cork for escape of the hot air.

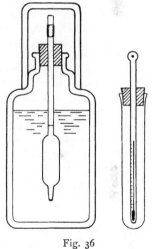

Fig. 36

TWO THERMOMETERS are employed, a small bulb reading to $\frac{1}{2}$ or 1° C., and a large bulb, accurately checked and reading to $\frac{1}{10}°$.

A graduation mark is made on the test-tube with a diamond or file at 2 c.c. from bottom, and the oil is poured into the tube up to

[1] Parkes (*Analyst*, 1918, **43**, 126) suggests the employment of a proportion of butyric acid to prevent freezing.

this mark[1] (meniscus reading) at the temperature of boiling water. The acid is measured at 20—25° C.

1. See that the test-tube is quite clean and dry.
2. Run in 2 c.c. of the clear oil, heated to boiling water temperature. This is judged by the mark on the tube. (It is undesirable to pipette it.)
3. Add 2 c.c. of acetic acid from bottle by means of the pipette. Insert the *small bulb* thermometer with cork, and heat, whilst cautiously shaking over naked flame until the mixture *suddenly clears*.
   Avoid contact of the thermometer bulb with the wall of the test-tube.
4. Allow to cool whilst still shaking and note the temperature at the first sign of turbidity of the liquid.
5. Repeat determination, using the *large bulb* thermometer (graduated in $\frac{1}{10}$°), but this time place in water heated to 5° above the temperature found in (4), and shake until clear. Then set aside the tube, immersed in the hot water, and carefully note the exact temperature when the faintest signs of turbidity appear.
6. Ascertain independently the acid value of the oil (as oleic), page 123.
7. Make a similar test with the same acid, and the standard almond oil.
8. Make corrections as shown below (§ 8).

(*b*) Ethyl-Amyl-Alcohol Reagent.

This may be kept in an ether bottle, or in a well-corked or stoppered bottle. The small bulb thermometer only is used in the test.

1. Operate exactly as described under Acetic Acid, 1—4 above, using the reagent in place of the acid.
2. Repeat in same manner, and take mean of observations.
3. Empty contents of tube into small beaker, rinse tube and thermometer with hot alcohol, add a few drops of phenol phthalein solution and titrate with N/10 alkali (see Acid Value, p. 123).
4. Calculate the acidity in terms of oleic acid, making the necessary calculation of the weight by multiplying the volume of the oil used by the specific gravity of the oil at 99—100° C. (which need only be approximate).
5. Note corrections for acidity as indicated below.

§ 8. Calculations

Correction for Acidity.

The correction for acidity of the oil should first be made. Ascertain, by reference to the following tables, the correction for 1 per cent. acidity for the class of oil under test, multiply this by the acidity found, and *add* the result to the temperature figure obtained. This is the final figure required in the case of the alcohol reagent. In the case of acetic acid, a correction is also necessary for the strength of acid.

[1] The authors have shown that results are not affected by *slight* variations in the proportions of the oil and solvent.

Influence of Acidity of Oils on Turbidity Figures.
(Acetic Acid.)

Class of Oils	Typical Oil employed	Fall in Turbidity temperature. Degrees per 1 per cent. Acidity (as Oleic)
Marine...	Whale	1·90°
Drying	Linseed	1·85°
Semi-drying	Cottonseed	1·77°
Non-drying (except rape and castor)...	Almond	2·27°
Vegetable fats (except coconut group)	Palm	2·10°
Rape oil group	Rape	2·23°
Coconut oil and palm kernel ...	Coconut	1·73°
Animal fats (except butterfat) ...	Lard	2·15°
Milk fats	Butterfat	1·41°

Influence of Acidity of Oils on Turbidity Figures.
(Amyl-Ethyl-Alcohol Reagent.)

Class of Oils	Typical Oil employed	Fall in Turbidity temperature. Degrees per 1 per cent. Acidity (as Oleic)
Marine...	Whale	1·95°
Drying	Linseed	2·05°
Semi-drying	Cottonseed	2·03°
Non-drying (1)	Almond	2·07°
Non-drying (2)	Rape	1·61°
Vegetable fats (except coconut group)	Palm	1·72°
Coconut group	Coconut	2·01°
Animal fats (except butterfat) ...	Lard	2·13°
Butter and milk fats	Butterfat	1·54°

Note.—If the class of the oil is unknown, a figure of 2·0° may be allowed in all cases, but the acidity of the oil, for this test, should preferably be low.

CORRECTION FOR ACETIC ACID.

Use the following formula for correction for the acid :

$$V = t + (80 - t'),$$

where V = " true Valenta,"

t = temperature obtained with oil tested, corrected for acidity,

t' = temperature with standard oil and the same acid (correcting for acidity if necessary).

Note.—If the standard almond oil is not *neutral* the acidity must be ascertained, and the correction made in all cases (see below, § 9).

The authors prefer the final figure as the ratio,

$$\frac{V \times 10}{80},$$

as this shows that the necessary corrections have been made, which a temperature figure leaves open to doubt.

§ 9. **Examples**

(*a*) USING ACETIC ACID.

The acetic acid purchased was frozen and the crystals drained. It then gave a figure of 72° C. with almond oil.

The acidity of the almond oil was 1·5 per cent. (as oleic), therefore the corrected figure for the acetic acid is $72° + (1·5 \times 2·27) = 75·4°$.

A sample of *olive oil* was tested and gave, using the small bulb thermometer, 67° C., and with the final test 66·7°.

The acidity of the olive oil was found to be 2·2, and the necessary correction is therefore $2·2 \times 2·27$, which, being added to the figure, gives 71·7°.

According to the formula

$$V = t + (80° - t')$$

we get $\mathbf{V} = 71·7 + (80° - 75·4) = \mathbf{76·3°}$,

and in the *ratio figure* we get

$$\frac{76·3 \times 10}{80} = \mathbf{9·54.}$$

(*b*) USING THE ALCOHOL REAGENT.

Industrial methylated spirit on being mixed with pure amyl alcohol gave a solvent showing a turbidity figure with almond oil of 55·5° C.

Acidity of almond oil (as oleic) = 1·2.

This oil is therefore standard at $70 - (1·2 \times 2·07)$ (correction for acidity of almond oil), i.e. 67·5°.

The amount of water required to raise the figure to 67·5 from 55·5 (12·0°) is $12 \times ·11$ (§ 5B) = 1·32 per cent.

A trial was made with 1·25 per cent. addition of water.

This gave 67·0°.

A further ·05 per cent. of water was added, giving the correct figure of 67·5° (equivalent to 70·0° for *neutral* almond oil).

A sample of linseed oil gave a turbidity figure of 58·3°.

On being titrated with N/10 alkali, 1·25 c.c. were used. Taking the S.G. of linseed oil at 99° C. as ·88, the acidity as oleic per cent. was

$$\frac{1·25 \times 5·61}{2 \times ·88 \times 2} = 2·0.$$

The necessary correction is therefore $+(2 \times 2·05)$

$$= +4·1 = \mathbf{62·4° \, C.}$$

The following tables and diagrams give the results obtained by the authors with various samples of oils, fats, and waxes:

Corrected Turbidity Temperatures (Vegetable Oils).

Standard neutral almond oil = 80·0° C. acetic acid.
Amyl-ethyl-alcohol solvent = 70·0° C.

Oil	Acidity per cent. (as Oleic)	Valenta Number Acetic Acid [t]	Standard Test Almond Oil [t']	Adjustment Factor for Acidity	True Valenta $V = t+(80-t')$	$\frac{V \times 10}{80}$	Turbidity Temperature Amyl-Ethyl-Alcohol	Adjustment Factor for Acidity	True Value
Perilla ...	5·5	19·0°	73·5°	1·85°	35·7°	4·5	49·0°	2·05	60·3°
Linseed ...	2·0	39·5°	76·0°	1·85°	47·2°	5·9	58·3°	2·05	62·4°
Tung ...	0·9	38·3°	73·5°	1·85°	46·5°	5·8	74·0°	2·05	75·8°
Soya bean ...	1·0	61·5°	90·0°	1·85°	53·3°	6·7	65·0°	2·05	67·0°
Niger ...	1·2	45·3°	73·5°	1·85°	54·0°	6·7	57·5°	2·05	60·0°
Sunflower ...	2·2	52·7°	73·5°	1·85°	63·2°	7·9	59·5°	2·05	64·0°
Maize ...	2·8	73·2°	90·0°	1·77°	68·2°	8·5	62·5°	2·03	68·2°
Cotton ...	0·1	69·5°	90·0°	1·77°	59·7°	7·5	65·0°	2·03	65·2°
Sesamé ...	4·0	73·0°	90·0°	1·77°	70·0°	8·7	60·5°	2·03	68·1°
Rape... ...	0·6	109·0°	80·0°	2·23°	110·3°	13·8	82·3°	1·61	83·3°
Almond ...	0·9	78·3°	80·0°	2·27°	80·3°	10·0	68·2°	2·07	70·1°
Arachis ...	1·1	95·8°	90·0°	2·27°	88·3°	11·0	72·0°	2·07	74·3°
Olive (1) ...	0·7	70·0°	80·0°	2·27°	71·6°	9·0	67·8°	2·07	69·2°
Olive (2) ...	1·8	72·5°	80·0°	2·27°	76·5°	9·5	65·5°	2·07	69·2°
Olive (3) ...	3·6	68·0°	80·0°	2·27°	76·2°	9·5	61·5°	2·07	69·0°

Corrected Turbidity Temperatures (Fats).

Standard neutral almond oil = 80·0° C. acetic acid.
Amyl-ethyl-alcohol solvent = 70·0° C.

Fat	Acidity per cent. (as Oleic)	Valenta Number Acetic Acid [t]	Standard Test Almond Oil [t']	Adjustment Factor for Acidity	True Valenta $V = t+(80-t')$	$\frac{V \times 10}{80}$	Turbidity Temperature Amyl-Ethyl-Alcohol	Adjustment Factor for Acidity	True Value
Cacao butter	2·9	88·0°	80·0°	2·10°	94·0°	11·7	71·0°	1·72	76·0°
Chinese vegetable tallow	6·9	69·8°	90·0°	2·10°	74·3°	9·3	54·0°	1·72	65·9°
Palm oil ...	0·1	89·8°	76·5°	2·10°	93·5°	11·7	68·0°	1·72	68·2°
Japan wax* ...	18·7	37·0°	88°	2·10°	68·?	—	44·0°	1·72	76·1°
Lard... ...	0·9	82·0°	76·5°	2·15°	87·5°	10·9	70·8°	2·13	72·7°
Tallow ...	0·1	102·0°	90·0°	2·15°	92·2°	11·5	72·5°	2·13	72·7°
Butterfat ...	1·9	45·3°	90·0°	1·41°	38·0°	4·7	43·0°	1·54	46·0°
Coconut ...	0·0	13·5°	81·5°	1·73°	12·0°	1·5	34·0°	2·01	34·0°
Palm kernel...	0·0	25·5°	81·5°	1·73°	24·0°	3·0	40·0°	2·01	40·0°
Coconut (2) ...	1·6	—	—	—	—	—	30·0°	2·01	33·2°

* Acidity too high for calculation.

Diagram of Turbidity Temperatures °C. (corrected).
Alcohol reagent.

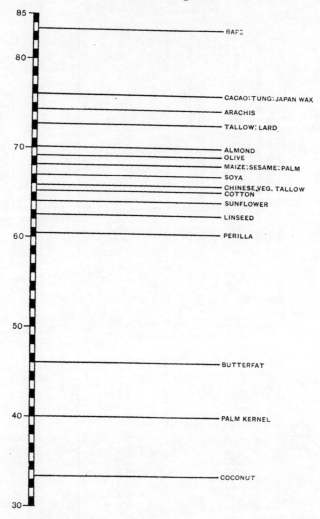

Diagram of Acetic Acid Turbidity (True Valenta)
numbers $\left[\dfrac{V \times 10}{80}\right]$.

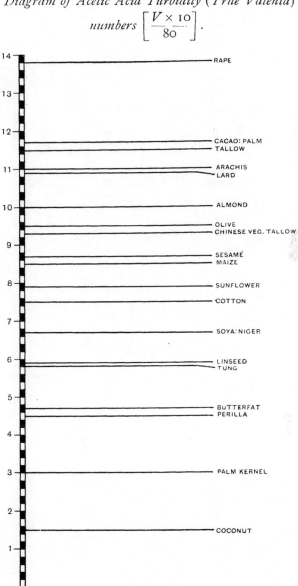

Corrected Turbidity Temperatures (Waxes).

Amyl-ethyl-alcohol reagent $= 70 \cdot 0°$ C. almond oil.

Wax	Valenta Number Acetic	Standard Test Acetic	True Valenta	$\dfrac{V \times 10}{80}$	Turbidity Temperature Amyl-Ethyl-Alcohol	True Value
Carnaüba	insoluble	61°	insoluble	—	82°	**82°**
Candelilla	,,	80°	,,	—	63°	**63°**
Beeswax	,,	80°	,,	—	76°	**76°**
Spermaceti ...	61°	61°	80°	10·0	44°	**44°**
Insect wax... ...	insoluble	61°	insoluble	—	insoluble	insoluble
Montan refined ...	68° [= S. P.]	61°	87°	10·9	70°	**70°**
Ozokerit	insoluble	61°	insoluble	—	insoluble	insoluble
Paraffin	,,	61°	,,	—	,,	,,
Beeswax + 10 per c. ceresin	,,	61°	,,	—	82°	**82°**
Candelilla + 10 per c. paraffin	,,	61°	,,	—	72°	**72°**
Candelilla + 33 per c. paraffin	,,	61°	,,	—	83°	**83°**

ROTATORY POWER   (7)

§ 1. Definition

ROTATORY POWER, or OPTICAL ACTIVITY, indicates the property which certain substances possess of rotating the plane of polarised light. The degree of activity of a liquid is determined by ascertaining the angle of rotation produced in passing a ray of polarised monochromatic light through a column of the liquid of known length.

§ 2. Remarks

In ordinary light, from whatever source obtained, the vibrations occur in all planes, and the light is capable of being refracted or reflected in any position. On passing light through certain substances, notably when of crystalline structure, the direction of the vibrations in the ray is confined to *one plane*. The ray of light is then known as POLARISED. Substances which possess optical activity have the power of turning the plane of polarised light either to the left (laevo-rotatory) or right (dextro-rotatory) to a definite extent, the degree of course varying according to the distance which the polarised ray passes through them.

The *Specific Rotatory Power* is defined as the angular rotation of the plane of polarised light by a column of liquid one decimetre in length, and of unit density. In the case of a homogeneous liquid it is given by *Biot's* formula

$$(a) = \frac{a}{l \times d},$$

where (a) = specific rotatory power,
            a = observed angular rotation in polarimeter,
            l = length of column of liquid in decimetres,
            d = density of the liquid.

For solutions, the formula becomes

$$a = \frac{100\,a}{l \times c},$$

where c = number of grammes of solute in 100 c.c. of the solution.

As polarised light of different wave-lengths is rotated to a different degree by optically active substances it is necessary to employ for determinations a monochromatic light of agreed wave-length. The light always used is that from a sodium lamp giving the D ray of the solar spectrum (wave-length = ·0005888 mm.). The other essential in the determination is a uniform temperature of the liquid during observation. In giving the Specific Rotatory Power of a substance, these two factors are represented as follows: $\left[a\right]_{D}^{20°}$, where the figure on the right hand above refers to the temperature of the observation in degrees C., and the lower letter to the character of the light ray employed.

§ 3. **Polarimeters**

In the construction of polarimeters, use is made of the crystal known as "Iceland Spar." This substance, as is well known, has the property of doubly refracting light. The crystal is cut in two, and the cut surfaces polished and cemented by Canada balsam. (In this way, one of the refracted rays of light is caused to be totally reflected.) The crystal so prepared is now mounted in a brass cell so that the light passes from end to end of the crystal, and, as before stated, one of the rays (the more highly refracted one) is totally reflected out of the field of view. As so designed, this is termed a *Nicol prism*. The emergent ray from the prism is polarised, and when two prisms are placed end to end, and a column of liquid is placed between the prisms, light being projected through the system, it will be found that, in a certain position of the prisms, no light is transmitted. On now rotating one or other of the prisms, light gradually appears until a maximum is reached, which occurs on rotating the prism exactly 90° (a right angle) from its original position. If this same prism be now rotated through a further angle of 90°, the field of view will again appear dark.

The reason for this is easily seen in the accompanying figure.

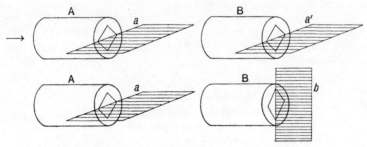

Fig. 37

A and **B** are the two prisms. The polarised light, on issuing from **A** is in the horizontal plane (*a*). On passing through **B**, the prism allows the whole of the light to be transmitted because it transmits light in the same plane (*a'*).

In the other case, when prism **B** is turned through a right angle, none of the light from prism **A** can become transmitted since **B** is now transmitting light in a vertical plane.

If the two prisms are now placed on a stand, and adjusted so that no light is visible, and a transparent, optically active substance is placed between them, the plane of polarisation is twisted round, and more or less light becomes visible. If the position of the front prism (termed the

" *analyser* ") is known, and this is now rotated until absolute darkness is restored, the back prism, or *polariser* remaining fixed, the angle through which it has been turned will be a measure of the optical activity of the substance in question.

Various forms of **Polarimeters** have been devised. In order to secure closer adjustment of the *analyser* the field of view is divided into two or three segments by means of discs of quartz or of other crystals. In this manner it has been possible to obtain an adjustment in which, for a given position of the analyser, the different segments of the field of view become of an identical shade of light, while the slightest rotation of the prism causes a difference of shade between the divisions of the field. In the *Laurent*, and the *Schmidt and Haensch* instruments, the field is divided into two equal segments, while in the *Lippich* polarimeter the field is divided into three. The latter instrument is here described, but all are very similar in manipulation.

§ 4. Application to Oils

All natural oils are optically active to a very small degree owing to the sterols which they contain. The figures obtained for the latter are as follows :

Cholesterol $\quad [a]_D^{15} = -31\cdot12$ Hesse.

Isocholesterol $\quad [a]_D^{17} = +59\cdot1$ Mareschi.

Phytosterols :

Sitosterol $\quad [a]_D^{15} = -23\cdot14$ Menzzi.

Brassicasterol $[a]_D^{18} = -64\cdot4$ Matthes and Heinz.

Sigmasterol $\quad [a]_D^{41} = -45\cdot01$,, ,,

In the case of sesamé oil, its activity ($+\cdot8$ to $+1\cdot6$) is due to the sesamin it contains (sesamin $[a]_D^{15} = +68\cdot3$ in chloroform).

With regard to castor oil, the glycerides of ricinoleic acid account for its high activity. (This is due to the asymmetric carbon atom of ricinoleic acid ; see Vol. I, page 51.) Castor oil $[a]_D^{15} +7\cdot6$ to $9\cdot7$.

There is also a group of oils, of less importance, the *Chaulmoogra* group, of which the glycerides are also optically active :

Chaulmoogra oil $\qquad [a]_D^{15} = +52\cdot0.$

Hyndocarpus (" marotti ") oil $\quad +56\cdot2.$

Lukabro oil $\qquad +42\cdot5.$

§ 5. The Lippich Polarimeter

The diagram, fig. 38, gives the arrangement of the optical parts of the instrument, and the photograph, fig. 39, is lettered correspondingly.

a. Glass cell containing solution of potassium dichromate.

b, c, d. System of lenses to give enlarged image of opening **e.**

n_1, n_2, n_3. Three Nicol prisms ; the plane of vibration of light issuing from n_2 and n_3 forms a small angle with that of n_1 which can be altered by arm. See fig. 39.

n_4. Analysing Nicol. When the scale is at zero, the plane of n_4 is

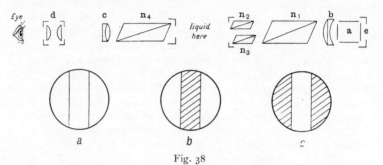

Fig. 38

Fig. 39. Lippich Polarimeter

so arranged that amounts of light from n_2, n_3 and n_1 are equal, and the field is uniformly lighted (a). In other positions the field is as figures a, b and c.

§ 6. **Method of Use**

1. Set up a source of monochromatic light (D ray) obtained by placing a piece of fused salt or borax in the special burner provided at 5 to 8 inches from the polarising end of the instrument: adjust the eye-piece till a sharp image of the field is obtained and rotate the graduated circle till the field is perfectly uniform.

2. Read scale by means of reading magnifiers, and if the instrument is in adjustment this should be 0° on one vernier and 180° on the other.

3. Fill a 200 mm. tube with distilled water and place it in the apparatus (fig. 39).

4. Refocus eyepiece, and carefully obtain position of uniform field. Take readings, and if not at zero and 180° use the readings obtained as zero, and subtract or add the subsequent readings of the instrument to obtain the true angular deviation.

5. If the oil to be examined is not quite clear, try warming and filtering it in a test-tube by means of a wad of dry cotton wool pushed down the tube. If still not clear, proceed as in (8). If clear, fill a dry 100 mm. or 200 mm. tube, place in instrument, refocus eyepiece and rotate graduated disc till a uniform field is again obtained. Note whether rotation is to right or left (clockwise or anti-clockwise).

6. Take readings; displace disc, and adjust again and take fresh reading. Repeat twice and reject inconsistent readings, averaging the others

7. Determine the temperature and density of oil, and calculate the Specific Rotatory Power

$$\left[a\right]_D^{t'} = \frac{a}{l \times d} \text{ (see page 87).}$$

8. If oil is too opaque to be used directly, dissolve in chloroform or acetic acid. Weigh out 10—20 grms accurately in a 50 c.c. graduated flask, add a little solvent, and dissolve and then make up to mark with solvent: shake well. Place this solution in a 200 mm. tube and observe the rotation as before. Calculate as follows:

$$\left[a\right]_D^{t°} = \frac{100\,a}{l \times c}.$$

9. If the rotation was to the *right*, the oil is *dextro-rotatory* (represented by a + sign), if to left it is *laevo-rotatory* (represented by a − sign).

Students should clean out the observation tubes when finished with, using warm soft soap solution and taking care to dry the screw caps that fit on either end, being particularly careful to avoid scratching the cover glasses.

THE IODINE VALUE ⑧

§ 1. Definition

The Iodine Value is the percentage of IODINE CHLORIDE (ICl) expressed in terms of Iodine which is absorbed by an oil, fat, etc.

§ 2. Outline of Method

An accurately weighed quantity of the oil, fat, etc., is dissolved in a suitable solvent, an excess of iodine chloride solution added, and after the requisite time the amount of ICl remaining unabsorbed is estimated, conveniently by the use of potassium iodide and thiosulphate, employing starch solution as an indicator. The difference between the original iodine content of the solution and its final content of iodine represents in terms of iodine the ICl absorbed by the oil, fat, etc.

§ 3. Remarks

A discussion of the theoretical and chemical, and also of the technical aspects of the Iodine Value will be found in Vol. I, p. 79.

Bromine was the halogen at first employed, but it was found that not only was it absorbed by the unsaturated carbon atoms, but it also replaced hydrogen by *substitution*, liberating HBr:

$$Br_2 + H - \overset{|}{\underset{|}{C}} - \text{ gives } Br - \overset{|}{\underset{|}{C}} - + HBr.$$

Iodine itself is only slowly absorbed, and cannot therefore be employed in practice.

The procedure at first used was due to *Hübl*, who mixed alcoholic solutions of iodine and mercuric chloride. These solutions interact with the production of iodine monochloride[1] as follows:

$$HgCl_2 + 2I_2 = HgI_2 + 2ICl.$$

All the older numbers for the iodine value were obtained by this method.

The **Wijs'** method is now almost exclusively employed. It consists in the use of a solution of ICl in glacial acetic acid, which, unlike the *Hübl* mercury solution, keeps stable for many months. The *Wijs* solution has the further great advantage that from it the ICl is absorbed very rapidly by the oil, fat, etc., so that, whereas the interaction formerly took several hours, it can now be completed, in most cases, within half an hour.

---

[1] In addition the following substances are also produced : HCl, HIO and HIO_3. For full details of the Hübl method see Lewkowitsch, *Oils, Fats and Waxes*, Vol. I. pp. 398 *et seq.*

Careful attention to the following points is necessary to ensure accuracy :

(*a*) The reagents used should be of the highest obtainable degree of *purity*.

(*b*) The vessels used must be scrupulously clean and *dry*.

(*c*) The iodine solution must be carefully measured (counting drops from pipette after the liquid has run out, and using same pipette for blank determination).

(*d*) The iodine solution should always be measured out at the *same temperature*.

(*e*) The titration should be carried out so that the end point is correct within ½ *drop* of the thiosulphate solution.

§ 4. **Materials required**

[A] Purest glacial acetic acid[1]

[B] Iodine   } or iodine trichloride.

[C] Bleaching powder and sulphuric acid }

[D] Potassium dichromate, purest and free from sodium.

[E] Starch.

[F] Sodium thiosulphate, pure crystals ($Na_2S_2O_3\,5H_2O$).

[G] Potassium iodide, purest (KI) 10 °/₀ solution.

[H] Carbon tetrachloride, redistilled and dried over fused calcium chloride.

§ 5. **Preparatory Operations**

I. PREPARATION OF IODINE MONOCHLORIDE (ICl) SOLUTION (WIJS' SOLUTION)

(i) Weigh out 13 grms of iodine[2] in a weighing bottle and dissolve in 1000 c.c. of purest glacial acetic acid, warming the solution on the water-bath. (Carefully avoid contact of the acetic acid with steam or vapour.)

(ii) Cool; pour about 100 c.c. into another vessel and set aside.

(iii) Fit up a chlorine apparatus.

Concentrated hydrochloric acid is allowed to drip on to a mixture of bleaching powder, with five times its weight of water. . The bleaching powder should be mixed into a paste with twice its weight of plaster of Paris and water, the paste allowed to set, and broken up into lumps. The chlorine gas is washed with a little water and then dried by bubbling through sulphuric acid in a wash bottle.

(iv) Pass the dry chlorine gas into the iodine solution until the dark brown colour of the liquid is seen to change to a *clear dark orange tint*. The iodine solution kept back (ii) is now added until the colour of the solution becomes faintly brown.

---

[1] Best if recently frozen, and any liquid poured off crystals.

[2] An alternative method is to dissolve a 10 gramme tube of iodine trichloride (as now prepared) in about 300 c.c. glacial acetic acid, and to add a 2½ per cent. solution of iodine in the same solvent until the iodine is slightly in excess (see I. iv). It is then made up to a litre with acetic acid.

This slight excess of iodine prevents the possibility of the formation of *iodine trichloride*.

(v) Heat the solution on the water-bath for 20 minutes. This renders it more stable, and it will not change in strength for several months. Place in a well-stoppered bottle and keep in the dark.

II. PREPARATION OF DECINORMAL (N/10) SODIUM THIOSULPHATE

(i) Weigh out 25 grms of pure crystallised thiosulphate ("hyposulphite") and dissolve in a litre of water.

12·5 grms in 500 c.c. water are sufficient for a few tests, as the solution will not keep for more than a few days.

(ii) Weigh out with great accuracy 4·9033 grms[1] of purest potassium dichromate, dissolve in distilled water and make up to 1000 c.c. This solution keeps indefinitely.

See that the water is at 15·5° C., and always use the solution at this temperature.

Place in bottle and label N/10 POTASSIUM DICHROMATE.

(iii) Pipette into a flask 10 c.c. of KI solution (10 °/₀), add 50 c.c. water and 5 c.c. of pure conc. hydrochloric acid. Then run in from a burette exactly 20 c.c. of N/10 dichromate.

(iv) Now titrate this iodine solution[2] with the thiosulphate solution (i), running in the thio. until solution is pale yellow, and then adding 10 drops of the starch solution (see below). Titrate very carefully until the blue colour is *just discharged* and gives place to a pale green coloration. The correct amount of water is then added to the thiosulphate solution to make it exactly equivalent to the dichromate[3]. (20 c.c. N/10 dichromate should take exactly 20 c.c. N/10 thio., see page 13.)

III. STARCH SOLUTION

Make a thin paste by rubbing up 1 grm of starch with a little water, and pour into 100 c.c. of boiling water. Heat for a few minutes on the water-bath till quite clear. The solution should be made up fresh each day.

IV. CARBON TETRACHLORIDE

This should have been freed from traces of moisture by placing a few granules of fused calcium chloride in the liquid.

V. THE SAMPLE

If this is an oil see that it is quite clear. If not, warm till clear, or, if still cloudy, filter through thick dry filter paper. If a fat or wax, melt and filter through a hot funnel, if necessary.

---

[1] Calculated from the equation

$$K_2Cr_2O_7 + 14HCl + 6KI = Cr_2Cl_6 + 8KCl + 7H_2O + 3I_2,$$

$K = 39·10$; $Cr = 52·0$; $O = 16·00$. Equivalent $K_2Cr_2O_7 = \dfrac{294·20}{6} = 49·033$.

[2] See equation above.
[3] Or the factor is noted.

§ 6. Procedure

1. If the character of the oil is not known, a preliminary test is advisable, as it is essential to have at least double the quantity of ICl solution than is actually required for the absorption.

   The simplest preliminary test is the bromine thermal test (page 97). The number of degrees rise × 5·5 will give the approximate iodine value.

2. Divide 10 grms by the approximate iodine value of the substance and weigh out very accurately about this quantity of the sample in a well-stoppered perfectly clean and dry flask of about 200 c.c. capacity.

   The weighing is best done as follows: Insert stopper, and weigh empty flask very accurately to ½ milligramme. Then melt the sample, if necessary, and with a graduated pipette run in an amount in c.c. corresponding to the number of grammes required × 1·1. This will give a suitable weight. Now ascertain very accurately the weight of flask and oil. Subtract weight of empty flask. This gives net weight of oil.

3. Dissolve in 5 c.c. of carbon tetrachloride.

4. Note the temperature of the Wijs' solution. (20° C. is a suitable working temperature.) By means of a pipette transfer 10 c.c. to the flask, counting the number of drops from the pipette after emptying.

5. Insert the stopper (moistened with a little KI solution) and gently mix contents of flask with a rotary motion. Allow to stand for 30 minutes.

6. Add 5 c.c. of the KI solution, and 50 c.c. water, and titrate with the standard N/10 thiosulphate as directed in § 5 (iv) above. (There is however no *green* coloration of the final liquid.) Note the number of c.c. of N/10 thio. used (*a*).

   Take no notice of the return of a blue tint after the titration is finished. This is due to decomposition of the fatty iodine product.

7. Make a blank determination (without the oil, etc.) using exactly the same quantity of the CCl$_4$, and pipetting the Wijs' solution in the same manner (counting the drops as before) at the same temperature. Allow to stand for the same time, and titrate with the N/10 thiosulphate after addition of KI solution and water. Note number of c.c. of thio. used (*β*).

Calculation

$$\text{Iodine value} = \frac{(\beta - a) \times \cdot 012692^* \times 100}{\text{weight of oil, etc. taken}} \ (\times \text{factor}).$$

Example

(i) Number of grammes of oil taken = 0·1125. (Iodine value about 100: = $\frac{10}{100}$ = 0·1 grm = 0·11 c.c. to be taken.)

---

* ·012692 = equivalent weight of iodine ÷ 10,000.

(ii)   20 c.c. of potassium dichromate (N/10) took 19·75 c.c. of thiosulphate solution,

$$\therefore \text{ factor of thio. solution} = 1\text{·}0127.$$

(iii)   Number of c.c. thio. used for test $(a) = 10\text{·}25$.

(iv)   Number of c.c. of thio. used for blank $(\beta) = 18\text{·}75$,

$$\therefore \text{ iodine value} = \frac{(18\text{·}75 - 10\text{·}25) \times \text{·}012692 \times 100}{0\text{·}1125} \times 1\text{·}0127$$

$$= \mathbf{97\text{·}1}.$$

Note.—If the oil belongs to the drying or semi-drying class, and great accuracy is required, it is best to take $2\frac{1}{2}$ times the weight of oil given, and use 25 c.c. of Wijs' solution. After the same lapse of time add 15 c.c. of KI solution, 100 c.c. water, and titrate as before.

As the Wijs' solution is quite stable for long periods, provided it is kept cool, and placed in a well-stoppered bottle, it is convenient to employ it as a standard when the strength has once been determined. This should be clearly marked on the bottle with the date of preparation and the dates of recheckings at intervals (using the standard dichromate) also recorded.

Thus, if 10 c.c. of the Wijs' solution are found to require 18·75 c.c. N/10 thiosulphate (§§ 6, 7 above), the strength of the solution (in terms of iodine) is **0·1875 N** (or **·023797 grms I per c.c.**). It thus becomes unnecessary after the first time to standardise the thiosulphate solution.

Students must leave all apparatus clean, rinsing out flask with hot soap solution, and cleansing burette by running water through. The thiosulphate solution need not be kept, but the N/10 dichromate should be put in a special bottle and reserved for this test only.

BROMINE THERMAL TEST   (8) a

Definition

This test is a measure of the heat evolved on treating a definite weight of an oil or fat dissolved in a definite volume of solvent with a definite volume of bromine.

Outline of Method

As usually[1] performed, 1 grm of the oil or fat is accurately weighed in a vacuum-jacketed test-tube, dissolved in 10 c.c. of chloroform, the temperature noted, and 1 c.c. of bromine at the same temperature rapidly added, the tube stirred with the thermometer, and the number of degrees C. rise in temperature recorded.

Remarks

During the absorption of bromine by unsaturated fats, heat is evolved proportional to the bromine absorbed.   The method thus furnishes a rapid means of indirectly determining the halogen absorption value of the fat (see Vol. I, p. 82).   It is not employed as much as formerly[2] since the iodine value, giving direct figures, is now capable of very rapid performance, using the *Wijs' iodine solution*.   The " Dewar " vacuum-jacketed tube is used as this reduces the heat lost by radiation to a minimum.   The figure obtained × 5·5 gives in many instances a fairly close approximation to the iodine value, but it must not be relied upon in all cases.

Apparatus and Materials required

[A] A Dewar vacuum-jacketed test-tube of about 20 c.c. capacity with a wire loop for suspension.

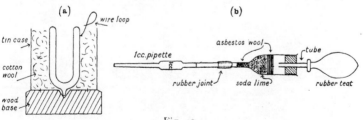

Fig. 40

[1] For suggested modifications see Wiley, *Jour. Soc. Chem. Ind.* 1896, 384 ; Archbutt, *Ibid.* 1897, 310 ; Heiduschka and Rheinberger, *Pharm. Zentralhalle*, 1912, 303.
[2] The authors have however found it of great use as a sorting test for samples of olive oil.

This is suitably placed in a small tin, with a hole punched in the centre of the bottom, standing on a wooden base, having a corresponding central hole to receive and protect the sealed end of the tube. The tube is packed round with cotton wool as in fig. 40 (a).

[B] A 1 c.c. pipette fitted with a rubber teat, and a soda lime protector as in fig. 40 (b).

[C] A thermometer graduated in $\frac{1}{5}°$ C.

[D] Pure anhydrous bromine.

[E] Anhydrous chloroform.

Procedure

1. Suspend the clean and dry vacuum-tube by its wire loop from the hook of the balance, and counterpoise.

2. Add 1 grm weight, and, by means of a fine piece of tubing, weigh in exactly this amount of oil or melted fat.

   Any excess may conveniently be removed by a spill of filter paper.

Fig. 41

3. Place tube carefully in nest, and add 10 c.c. of chloroform. Stir gently with thermometer, and note temperature (°C.) accurately (a).

4. See that the bromine is at same temperature, and then draw up to mark of pipette by inserting and squeezing rubber ball, and allowing to fill.

5. Run bromine into solution meanwhile stirring carefully with thermometer.   Watch mercury and note top point of rise (β).

6. **Thermal value** = final temperature (β) – initial temperature (a).

   APPROXIMATE IODINE VALUE = THERMAL VALUE × **5·5**.

Students should rinse out the Dewar tube with a little chloroform and replace in nest.   The bromine pipette need not be cleansed.   Replace thermometer in case.

OXYGEN ABSORPTION TEST ⑧ c

Definition

The test aims at recording the percentage of oxygen capable of absorption by the oil under ordinary atmospheric conditions.

Remarks

Although the "drying" power of oils on exposure to air has been adopted as the basis of classification of oils, no satisfactory method has been found to determine quantitatively this drying power. "Drying" is undoubtedly due mainly to oxygen absorption from the air, but the rate of such absorption is dependent on many factors which it is difficult to control sufficiently carefully for analytical purposes.

The iodine absorption value has therefore, in analytical work, been taken as a measure of the oxygen absorptive power of oils.

One of the main reasons why concordant results have not been obtained in the past is because it is very difficult to have a layer of oil sufficiently thin for the purposes of the test. Thus the external surfaces of the oil become oxidised, and the skin formed protects the interior from oxidation.

Livache distributed the oil over finely divided lead, and *Lippert* used copper for the purpose.

Take 1 grm of lead powder (obtained by pptg. a lead salt with zinc, washing ppt. with water, alcohol and ether and drying in a vacuum), weigh it accurately on a watch glass, and spread it in a thin layer. Drop from a pipette ·6—·7 grm of the oil on to the powder as evenly as possible. Stand watch glass at ordinary temperature exposed to light, and ascertain increase of weight from time to time.

Bishop adds manganese rosinate and spreads the oil on precipitated silica.

Fahrion used a strip of chamois leather for distribution of the oil. The authors prefer fine wire gauze, as this does not tend to absorb moisture (one of the chief difficulties), and a thin film of oil is formed in the interstices of the gauze, which is exposed on both sides to the action of the air.

In any case, the information obtainable, though of great interest, is inferior in value to that yielded by the iodine value.

An important technical problem in the paint and varnish trades, in addition to the oxygen absorption, is the character of the film produced, especially its elasticity, transparence and lustre.

For this purpose the oil, suitably treated (boiled, plus driers, etc.), is painted on wood, glass, or preferably on a piece of gelatine film. The transparence, and especially the elasticity, can then be tested on the drying of the oil, and if further tests are desired, the gelatine may be dissolved away by immersion in warm water and the film of dried oil tested in any suitable manner.

Procedure. (F. and W. modification.)

1. Take a piece of fine copper gauze, preferably about 160 mesh to inch, and cut a strip 1½ inches square, roll the strip into cylindrical form. (See fig. 43.)

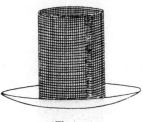

Fig. 43

2. Weigh accurately on a watch glass. Then dip the strip in the oil to be tested and allow to drain on a filter paper for a few minutes.
3. Place on the watch glass and weigh accurately. Usually about 100 milligrammes of oil will be retained by the gauze.
4. Place the gauze and watch glass preferably in an incubator at about 20—22° C. in the dark and weigh each day, recording increase of weight. (If an incubator is not available, a dark cupboard will serve.) In order to control the moisture content of the air, a large dish of water should be placed in the incubator and renewed from time to time as it evaporates.

    If desired, the oil may previously be treated with a drier or "boiled," and compared with an untreated control.

ELAÏDIN TEST ⑨

§ 1. Definition

The Elaïdin Test gives an indication of the amount of the glyceride of *oleic acid* present in an oil, since, on treatment with nitrous acid, oils containing notable proportions of olein are converted into solid or semi-solid bodies, owing to the production of the solid stereo-isomer ELAÏDIN.

§ 2. Outline of Method

The oil is added to a solution of nitrous acid, prepared by the action of mercury on nitric acid, and the two are well shaken together at intervals, the character of the resulting product being then noted.

§ 3. Remarks

With regard to the phenomenon of *stereo-isomerism*, see Vol. I, pages 44—46. The elaïdin test is not of great value, and is qualitative only. It has been shown[1] that the results obtained are influenced by many different factors, as e.g. the method of preparing the nitrous acid, of mixing the acid and oil, the temperature, and the age and condition[2] of the oil. The test is chiefly used for olive oils, and in this case a comparative test should always be made at the same time with a sample of olive oil of known purity.

§ 4. Apparatus and Materials required

[A] Wide-mouthed stoppered bottles of about 100 c.c. capacity.

[B] Pipettes ; 2 c.c., 50 c.c.

[C] Nitrous acid solution.

Measure out 1 c.c. of mercury into a dry beaker, add 12 c.c. of nitric acid of s.G. 1·42 and allow to act in the cold till all the mercury is dissolved. The nitrous acid formed is absorbed with the production of a green solution: as long as the green coloration remains the solution is fit for use.

§ 5. Procedure

1. Into the wide-mouthed bottle [A] pipette 50 c.c. of the oil to be tested.

2. Add 2 c.c. of the nitrous acid reagent [C], insert stopper and well shake.

3. Place the bottle in water maintained at a temperature of 25—30° C., and re-shake at intervals of 10 minutes for 2 hours.

4. Note the character of the product obtained ; if solidification occurs, note also the time taken for this.

[1] *Hübl*; *Welleman.*

[2] *Gintl* showed that after exposure to sunlight for 14 days, olive oil did not yield an elaïdin.

Results

The following classes of products may be distinguished with the different oils :

Olive oil Almond Arachis Lard Sperm Neat's foot (sometimes)	solid hard mass.
Neat's foot oil Arctic sperm Mustard Arachis ⎫ Sperm ⎬ (sometimes) Rape ⎭	buttery mass.
Sunflower oil Niger Soya bean Cottonseed Sesamé Rape Cod liver Seal Whale Porpoise	pasty mass, separating from a fluid portion.
Linseed oil Hempseed Walnut	liquid products.

Students should leave any reagent over, in the cylinder, and label same. Bottles to be cleansed with warm soap solution, rinsed and dried.

SAPONIFICATION VALUE   ⑩

§ 1. Definition

The saponification value indicates the number of *milligrammes* of potassium hydroxide required to completely saponify *one gramme* of the substance. It can therefore be simply expressed :

sap. value = $^\circ/_\circ$ KOH for saponification × 10.

The SAPONIFICATION EQUIVALENT is the number of grammes of oil saponified by one equivalent (56·1 grms) of potash.

$$\therefore \text{ sap. value} = \frac{56,100}{\text{sap. equivalent}},$$

$$\text{sap. equivalent} = \frac{56,100}{\text{sap. value}}.$$

§ 2. Outline of Method

The fat etc. is boiled for half an hour with an excess of alcoholic potash, the amount of unabsorbed potash being then determined by back-titration with N/2 HCl ; a blank determination is also made, and the sap. value is calculated from the difference between these two figures.

§ 3. Remarks

For the practical applications of the saponification value see Vol. I, pages 83—85. In the case of WAXES the flask should be boiled for an hour or longer, and a solution of sodium in alcohol (sodium alcoholate) is preferably employed.

In the case of VERY DARK SUBSTANCES (as e.g. the crude bitumen waxes) the solution must be diluted with alcohol after saponification until the colour change with the phenol phthalein can be distinctly seen, and a large porcelain basin may be employed for the titration.

§ 4. Apparatus and Materials required

[A] 2 flasks, 200 c.c. capacity.

> These are preferably round-bottomed Jena flasks (in fig. 44 a conical flask is shown), and in any case should be of the same glass, as the potash has a pronounced solvent action on some kinds of glass employed.

[B] 2 condensers on stand (upright "reflux" as at **b**).

> A convenient stand for boiling under reflux is that shown in the photograph, fig. 48 on page 121, which can be made by a tinman for a few shillings. It is made of soldered tin-plate with a wood base, and ensures that the blank test is performed under exactly similar conditions with the determination. (A water-bath, as in fig. 44, may replace the sand-tray here shown, but is not necessary.)

[C] 25 c.c., 10 c.c., and 2 c.c. pipettes, and 50 c.c. burette.

[D] Alcoholic potash.

> This is prepared by dissolving about 20 grms of potash (purified from alcohol) in the minimum quantity of water, and making up to 500 c.c.

with alcohol. The solution is allowed to stand, and the clear portion can be poured off for use.

Fig. 44

Ordinary methylated spirit may be employed if treated in the following manner:

Take a litre of methylated spirit, and add powdered potassium permanganate until a dark purple coloration is obtained. Allow to stand

several hours in a warm place, adding more $KMnO_4$ if the purple coloration disappears. Then place a little powdered chalk in the flask, and
distil, testing the distillate till, when boiled in a test-tube with a little
strong KOH solution, no yellow coloration occurs on standing. Collect
the remainder of the distillate for use, but reject the last 50 c.c.

[E] N/2 HCl, accurately standardised. (N/2 H_2SO_4 is unsuitable.)

[F] 1 °/₀ phenol phthalein in alcohol.

§ 5. Procedure

1. Accurately weigh the clean dry flask (if round-bottomed suspend
   by wire from hook over balance pan) and then add 2 c.c. of the oil
   (or melted fat or wax) with a pipette, taking care that no oil adheres
   to neck of flask. Accurately weigh again.
2. Run in 25 c.c. of the alcoholic potash with pipette, counting the
   drops which are added after the pipette is emptied.
3. Place under reflux condenser. Then take the other flask, and run in
   25 c.c. of potash as before, allowing the same number of drops to
   fall from pipette. Attach to reflux.
4. Gently boil both flasks for 30 minutes, rotating the oil mixture
   occasionally to ensure mixing.
5. Disconnect latter, add 10 c.c. of carefully neutralised alcohol and
   titrate with N/2 HCl, using 1 c.c. of phenol phthalein. Note number
   of c.c. used (a).
6. Titrate blank determination in same manner after addition of
   10 c.c. alcohol. Note result (b).
7. **Saponification value**

$$= (b-a) \quad \times \quad \frac{56 \cdot 1}{1000 \times 2} \quad \times \quad \frac{100}{\text{wt. of fat}} \quad \times \quad 10,$$

c.c. N/2 KOH	wt. of KOH	convert to	milligrammes
absorbed	in 1 c.c. of	percentage	of KOH
	N/2 solution	of fat	per 1 grm fat

or simplified thus :

$$\frac{(b-a) \times 28 \cdot 05}{\text{weight of oil}}.$$

§ 6. Example

Wt. of oil taken = 1·8035 grms.

Volume of N/2 HCl required by blank = 35·85 c.c.

„   „   „   „   „ oil   = 24·05 c.c.

Hence volume of N/2 HCl required by KOH used (35·85 − 24·05) = 11·8 c.c.

1·8035 grms oil require 11·8 × ·02805 grm KOH.

∴ 1 grm oil requires $\dfrac{11 \cdot 8 \times 28 \cdot 05}{1 \cdot 8035}$ mgrms.

∴ Sap. value = 183·5.

Students should leave flasks clean, and N/2 HCl *in burette should be
emptied back into its bottle and the burette washed through with distilled
water.*

INSOLUBLE BROMIDE VALUE
OF FATTY ACIDS   (11)

§ 1. Definition

The INSOLUBLE BROMIDE VALUE of the fatty acids derived from an oil or fat is the percentage of insoluble bromides obtained on brominating a solution of the fatty acids in a suitable solvent.

§ 2. Outline of Method

A weighed quantity of the oil is saponified, the fatty acids separated, dissolved in ether with a little acetic acid, and chilled. The bromine is added drop by drop with control of the temperature, until in definite excess, and after standing for a minimum time, is filtered, washed and dried to constant weight.

§ 3. Remarks

The test was originally proposed by HEHNER AND MITCHELL in 1898, and was based upon the investigations of HAZURA just ten years earlier. The fatty acids from the oils were originally used, but Hehner and Mitchell preferred the oils themselves as they found the precipitated bromides less soluble. Their method was to take 1—2 grms of the oil and dissolve in 40 c.c. of ether, to which a "few c.c." of glacial acetic acid were added (they found a better character of precipitate using the acetic acid than with ether alone). The temperature was kept at 0° C. and several hours were allowed for precipitation. Filtration was carried out by means of a thistle funnel, covered with chamois leather, and a filter pump, and the precipitate was washed with four successive lots of 10 c.c. of ether at 0° C.

LEWKOWITSCH specified the temperature of 5° C. for operating, filtered through a plaited filter paper, and preferred to operate on the mixed fatty acids.

PROCTOR[1] brominated a solution of the oil in carbon tetrachloride, and precipitated the bromides by the addition of absolute alcohol.

SUTCLIFFE[2] suggested the use of 5 c.c. of glacial acetic acid with 40 c.c. of ether, and found much lower results if more acetic acid were used. He cooled at the temperature of tap-water (about 11° C.), and operated on the oil.

EIBNER AND MUGGENTHALER[3] worked out an elaborate method, operating on the fatty acids, and employing a temperature of − 10° C.

[1] *Jour. Soc. Chem. Ind.* 1906, **25**, 798.   [2] *Analyst,* 1914, **39**, 28.
[3] Abstracted in *Analyst,* 1913, **38**, 158.

The foregoing methods were carefully reviewed by GEMMELL[1] with a view to establishing a standard method of procedure, capable of yielding consistent results. He found that variations occurred in the original method of Hehner and Mitchell with the concentration of the solution of the oil, and the amount of acetic acid employed. Also that, using the oil, the bromine content of the precipitate was liable to variation. The minimum time for complete precipitation was found to be four hours. Difficulty was experienced with Eibner and Muggenthaler's method, as being too cumbersome, occupying an undue length of time (three days) and employing a temperature which was difficult to maintain (‑ 10° C.).

In the authors' opinion, the method outlined by Gemmell is the most satisfactory one and capable of yielding concordant results. The method here given therefore is broadly that of Gemmell with some modifications worked out by the authors chiefly with a view to shortening the time required.

In regard to chemical considerations see Vol. I, pp. 85, 86.

§ 4. Apparatus and Reagents required

[A] A double-bulbed flask (as used for butter analysis) as shown.
    This is provided with a blow-off device as employed in wash bottles. (Fig. 45.)

[B] 100 c.c. graduated flask, with stopper.

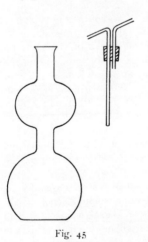

[C] Small flasks provided with accurately-ground stoppers of about 50 c.c. capacity (weighing bottles are suitable).

[D] A small centrifuge taking tubes of 10 c.c. capacity (or over).

[E] Rubber-tipped stirring rods.

[F] Swedish filter papers and funnels.

[G] Ether free from alcohol and water.
    Take 200 c.c. of ether in a 500 c.c. flask, add 10 grms of pure caustic soda, and allow to stand for two days. Then distil the ether over a water-bath (using due caution as regards naked flames) and to the distillate add 10 grms of anhydrous copper sulphate. Allow to stand several days.

Fig. 45

[H] Glycerin-caustic soda solution.
    Dissolve 100 grms or less of pure stick caustic soda in an equal weight of water. To every 20 c.c. of this solution add 180 c.c. of pure concentrated glycerin.

[I] Sulphuric acid 10 per cent. solution in distilled water.

[J] Glacial acetic acid.

[1] *Analyst*, 1914, **39**, 297.

[K] Bromine (pure).
[L] Ice.

§ 5. Procedure

1. Into the double-bulbed flask [A] weigh 5 grms of the oil to be tested.

2. Add 20 c.c. of glycerin-caustic soda solution [H], and cautiously heat over a naked flame for 2—3 minutes (rotating flask by a movement of the wrist) until the liquid has become clear.

3. Set aside to cool, and when still hot (but not over 100° C.) add 50 c.c. water, and rotate flask until the soap formed is dissolved.

4. Add from a pipette 20 c.c. of 10 °/₀ H_2SO_4 [I], and place in boiling water until the fatty acids are melted.

5. Cool under tap and add about 30 c.c. of ether [G]. Cork and shake until the fatty acids are dissolved, and allow the ether to separate.

6. Insert the cork and tubes [A], and carefully blow off the ether into a 100 c.c. graduated flask.

7. Add a further 20 c.c. of ether to flask, shake and settle, and blow off once more. Repeat with further 20 c.c. ether.

8. The combined ether solution is made up to 100 c.c. and shaken.

9. Of the above, 20 c.c.[1] is pipetted into a 50 c.c. stoppered flask [C], 2 c.c. of glacial acetic acid added, the stopper inserted and the flask cooled to 0° C. in ice and water.

10. Bromine is now added, slowly, drop by drop (preferably from a burette), rotating the flask and keeping it in the ice water, until a permanent red-brown coloration is obtained.

11. The flask is stoppered and allowed to stand in ice water *at least* four hours.

12. Decant the brown liquid cautiously from the deposited crystals as closely as possible, and transfer the crystals to the centrifuge tube [D]. Centrifuge for three minutes at moderate speed.

13. Pour off the ether from the tube and add 5 c.c. of chilled dry ether [G] to the flask, and by means of a rubber-tipped rod clean the sides of the flask and transfer to the centrifuge tube. Stir up the liquid and crystals thoroughly with a pointed stirring rod and re-centrifuge.

14. Pour off liquid and repeat mixing with further 5 c.c. chilled ether. Centrifuge and pour off ether.

15. Transfer crystals to a tared, dry filter paper with a further 5 c.c. of chilled ether. Dry to constant weight in water oven.

16. The weight of crystals found calculated to percentage of original oil = Insoluble Bromide Value (= weight of crystals × 100).

Note.—The m.p. of the crystals may be taken. Vegetable oils give insoluble bromides which melt to a clear liquid at about 175—180° C. Marine oils yield bromides which remain solid up to 200° C., and then darken and decompose.

[1] From the remaining 80 c.c. other check estimations may be made.

§ 6.   *Gemmell* gives the following figures :

Linseed oils (raw)	...	...	...	32·6 to 37·6 per cent.
,, (boiled)	...	...	...	25·9 to 33·9 ,,
Soya bean oil	...	...	...	4·10 ,,
Rape (1)	...	...	...	2·35 ,,
,, (2)	...	...	...	5·80 ,,
Walnut	...	...	...	3·00 ,,
Tung	...	...	...	nil ,,
Cod liver	...	...	...	35·20 ,,
Whale	...	...	...	21·70 ,,
,, (brown)	...	...	...	25·8 ,,
Menhaden	...	...	...	51·7 ,,
Shark liver	...	...	...	17·7 ,,
Seal	...	...	...	29·12 ,,
Sperm oil	...	...	...	1·70 ,,

*Note for students.   All ether washings to be placed in "Ether Residues"
bottle for recovery.   Cleanse all glass apparatus with hot soft soap solution
and rinse in water.*

REICHERT-MEISSL VALUE

(With Wollny's[1] and Leffmann and Beam's[2] modifications.)

§ 1. Definition

The Reichert-Meissl Value is a measure of the *soluble volatile fatty acids* (i.e. acids of low molecular weight) obtained from 5 grms of a fat or wax.   Uniform results are obtained providing the conditions laid down are strictly adhered to.   The figure is in terms of cubic centimetres of decinormal alkali required for neutralisation of these acids.

The figure for the soluble volatile acids obtained under the conditions of the Polenske test is practically identical with the above, but is termed the **Reichert-Polenske** for distinction.

§ 2. Outline of Method

The fat is saponified, and the resulting soap dissolved in water, the fatty acids liberated by the addition of sulphuric acid, and the liquid distilled, filtered, and filtrate titrated with N/10 alkali.

§ 3. Remarks

In *Reichert's*[3] original method 2·5 grms of the fat were taken but it was found more convenient to adopt *Meissl's* suggestion to double this quantity for the determination.

Wollny[1] then showed that unless certain conditions were laid down and strictly adhered to, uniformity in the results was not attainable. These were principally :

    (1) A definite size and shape of apparatus.

    (2) A fixed time for the distillation of a definite volume of distillate.

    (3) Prevention of dissolved carbon dioxide in the distillate.

    (4) Avoidance of agglomeration of unmelted fatty acids.

    (5) Reducing liability of ester formation[4] of the volatile acids.

A committee was appointed and standard conditions laid down for the test.   A further modification by *Leffmann* and *Beam*[2] was the use of a solution of caustic soda in glycerin, for saponification, in place of alcoholic potash, and as the method is thereby greatly shortened, and the results obtained are identical, the modified procedure is adopted here. [The accompanying photograph shows the distillation apparatus fixed up ready for use, and gives the standard dimensions.]

---

[1] *Jour. Soc. Chem. Ind.* 1887, 831, and *Analyst*, 1900, 309.
[2] *Analyst*, 1891, 153.
[3] *Zeits. f. analyt. Chem.* 1879, 1868.
[4] Lewkowitsch doubts the possibility of esterification. See *Oils, Fats and Waxes*, Vol. I, p. 418.

§ 4. **Solutions required**

[A] A solution of caustic soda in glycerin.
>Weigh out 50 grms caustic soda.
>Dissolve in 50 c.c. water.
>Mix 50 c.c. of this solution with 450 c.c. of pure glycerin.

[B] Normal sulphuric acid.

[C] Decinormal caustic soda accurately standardised (or barium hydrate).

[D] Phenol phthalein solution.
>Dissolve 1 grm phenol phthalein in 100 c.c. alcohol.

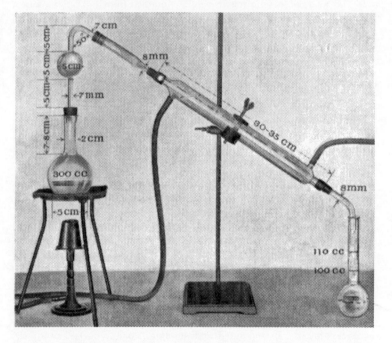

Fig. 46

§ 5. **Procedure**

1. See that the distillation flask (300 c.c.) is clean and dry.
2. Counterbalance the flask, and weigh into it exactly 5 grms of the filtered oil or fat.
3. Add 20 c.c. of solution [A] to flask with a pipette.
4. Heat cautiously over a naked flame (with constant rotary movement to prevent cracking and spurting) until liquid becomes perfectly clear (2—3 minutes).
5. Stand aside to cool. The temperature must be allowed to fall below 100° C.

6. Boil some distilled water for 10 minutes, and add 100 c.c. to flask. Shake till soap is dissolved.

7. Add 40 c.c. N. sulphuric acid, and a few fragments of pumice and connect at once to condenser, using naked flame and asbestos[1] provided.

8. Heat gently, avoiding boiling the liquid, until the fatty acids are quite melted.

9. Distil off 110 c.c. into the graduated flask during about 30 minutes.

10. Shake distillate, and filter 100 c.c. into a flat-bottomed flask.

11. Add $\frac{1}{2}$ c.c. phenol phthalein soln. [D] and titrate with N/10 NaOH [C] until pink coloration is just permanent.

12. The R.-M. value = no. of c.c. N/10 NaOH × 1·1.

It is advisable to make a blank estimation, using the same materials, i.e. heating [A] solution (but without oil) for the same time and then proceeding exactly as described. The no. of c.c. required of N/10 alkali should not exceed 0·3 c.c. This figure should be subtracted from the result obtained in (12), and should be fairly constant for the same materials.

§ 6. **Interpretation of Results**

See page 86 of Volume I of this book ; also under *butterfat*, page 173.

Since old and rancid oils undergo more or less decomposition the fatty acids of high molecular weight being split up into volatile acids, such specimens will give a high R.-M. value. In the fresh condition, the great majority of oils and fats (with the exception of those noted on page 87) give values not exceeding 1·0.

Example

A sample of edible fat consisted entirely of a mixture of coconut oil and butterfat. Assuming that the R.-M. value of the coconut oil was 7 and that of the butterfat 30, and the fat gave a R.-M. value of 13, what proportion of each was present in the mixture?

These were in the proportions of

$$(13 - 7) \text{ butterfat to } (30 - 13) \text{ coconut}$$
$$= 6 : 17 = \frac{6}{6 + 17} \times 100 \,°/_° \text{ butterfat,}$$
$$\frac{17}{6 + 17} \times 100 \,°/_° \text{ coconut}$$
$$= \begin{cases} \text{butterfat} & \dots 26 \\ \text{coconut oil} & \dots 74 \end{cases} 100.$$

Students should cleanse all apparatus in hot soft soap solution, and rinse in water.

---

[1] A piece of asbestos 12 cm. diameter with a hole in centre 5 cm. diameter.

THE POLENSKE VALUE (12) a

§ 1. Definition

The Polenske Value is a measure of the *insoluble* volatile fatty acids obtained from 5 grms of a fat or wax. It is an extension of the previous (R.-M.) test, but greater exactitude in its performance is necessary, and the prescribed directions must be minutely adhered to. The figure is in terms of cubic centimetres of decinormal alkali required for neutralisation of the fatty acids.

§ 2. Outline of Method

The fat is saponified, the resulting soap dissolved in water, the fatty acids liberated by the addition of sulphuric acid, the liquid distilled, the distillate filtered, and the insoluble acids dissolved in alcohol and titrated with $N/10$ alkali.

§ 3. Remarks

Adherence to exact conditions in the distillation (and subsequently) is of even more importance than in the Reichert-Meissl test, since the amount of insoluble acids carried over with the steam depends on the temperature of the liquid—local overheating increasing the amount,—the amount of steam produced in a given time, the area of surface of liquid exposed etc. The distilling flask, still head bulb and tube, and condenser must be also of standard dimensions, as given in photograph, which shows the apparatus filled ready for distillation.

The "REICHERT-POLENSKE," or titration figure for the soluble volatile acids, may, for most purposes, be taken as identical with the Reichert-Meissl. The KIRSCHNER and SHREWSBURY-KNAPP (ELSDON) tests are extensions of this determination. (See pages 147, 154.)

§ 4. Reagents required

[A] Caustic soda solution.   Made by dissolving pure sodium hydrate in an equal weight of water.

[B] Glycerin.

[C] Sulphuric acid.   25 c.c. of pure concentrated H_2SO_4 to 1000 c.c. with distilled water. (35 c.c. should just neutralise 2 c.c. of [A].)

[D] Alcohol.   90 per cent. carefully neutralised (phenol phthalein).

[E] Decinormal sodium hydrate or barium hydrate.

[F] Phenol phthalein solution.   1 per cent. in alcohol.

[G] Powdered pumice, sifted through "butter muslin" or passing "90 mesh."

§ 5. Procedure

1. See that the 300 c.c. distillation flask is clean and dry.

2. Counterbalance, and weigh into it exactly 5 grms of the filtered fat.

3. Add 2 c.c. of NaOH solution [A] and 20 grms glycerin.
4. Heat cautiously over a naked flame (with constant rotary motion to prevent cracking) until liquid becomes perfectly clear (2—3 minutes).
5. Set aside to cool.
6. When below 100° C., 100 c.c. of water are added, and solution of the soap effected; if necessary by warming on water-bath.

    The solution should be clear and pale straw coloured. If brown, the test must be rejected.

7. 40 c.c. of sulphuric acid solution [C] are added, and 0·1 grm of powdered pumice [G].
8. The flask is attached to condenser, and the fatty acids liquefied by a gentle heat, avoiding boiling the liquid. The small hole in the still head tube should be about 1 cm. from underside of cork.
9. The liquid is distilled, using a naked flame, and the asbestos plate[1] provided, and the flame regulated, so that 110 c.c. are distilled off within 19—20 minutes. Also, the condensed distillate must drip

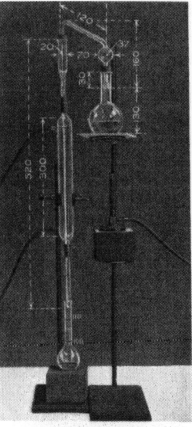

Fig. 47

into the receiver at a temperature not higher than 18—20° C. If higher than this a greater flow of cooling water must be employed.
10. Immediately 110 c.c. have distilled over, the flame is taken away, the receiving flask removed, and at once replaced by a 20 c.c. measuring cylinder.
11. The distillate in the flask is cooled by immersion up to the 110 c.c. mark in water at 15° C. for about 5 minutes.

[1] See footnote, page 115.

12. The floating acids (forming oily drops) are now made to adhere to the neck of the flask by gently tapping, and the flask immersed for a further ten minutes.

13. The insoluble acids are now examined, and a note made of whether they appear as (1) oily drops, (2) semi-solid, (3) solid.

14. Insert a cork into flask and reverse several times.

15. Filter through a filter paper of 8 cm. diameter. The titration of 100 c.c. of the filtrate, as directed in the previous test, gives the "Reichert-Polenske" figure.

16. Wash condenser tube, 20 c.c. cylinder, and measuring flask with 15 c.c. water three times, using same water for each vessel, and placing on filter. Reject wash waters.

17. Rinse each again with 15 c.c. alcohol three times and pour on filter. Each 15 c.c. must be allowed to drain through before the next is added.

18. Titrate alcoholic filtrate with N/10 alkali, using 0·5 c.c. phenol phthalein solution [F]. This figure = Polenske Value.

§ 6. Interpretation of Results

The test has reference only to coconut and palm kernel oils and butterfat.

Coconut and palm kernel oils, yielding more of the *intermediate* mol. wt. fatty acids on hydrolysis, give high numbers, while butterfat, the volatile acids of which are chiefly *low* mol. wt., gives low figures.

As regards the appearance of the insoluble acids (13 above) those from pure butterfat with low R.-M. values are usually liquid and clear, while those from fat of high R.-M. value are semi-solid and not clear. If over 10 per cent. of coconut oil is present, the acids are always liquid.

The matter is fully discussed in Vol. I, pages 78, 87, 173.

Example

A sample of butterfat gave the following figures :

Reichert-Meissl value 21·2,

Pol. value 3·7,

the insoluble volatile acids remained liquid on cooling (§ 5, 13 above). Was coconut oil present, and if so in what proportion?

Referring to the table given on page 175, Vol. I, we find that a genuine butter of R.-M. val. 21·2 should have a corresponding Pol. val. of 1·4. There is therefore an excess of (3·7 − 1·4) = 2·3 in the normal value.

By Polenske's calculation, each 0·1 c.c. is equivalent to 1 per cent. of added coconut oil.

This gives 23 per cent. added coconut oil present.

Students should cleanse all apparatus in hot soft soap solution, and rinse in water.

THE ACETYL VALUE   ⑬

§ 1. Definition

The acetyl value is the number of milligrammes of potassium hydroxide required to neutralise the acetic acid obtained when 1 gramme of an acetylated oil, fat or wax is saponified.

§ 2. Outline of Method

The fat etc. is boiled under a reflux condenser, with excess of acetic anhydride, the acetylated fat washed till free from acid and dried by filtration.   The product is then saponified with a measured excess of alcoholic potash, an equivalent quantity of standard sulphuric acid is added to the aqueous soap solution, the liberated fatty acids filtered off, and the acetic acid in the filtrate titrated with standard alkali.

§ 3. Remarks

As explained in Vol. I, page 87, the method gives a measure of the hydroxylated fatty acids free, or as glycerides in an oil or fat, and of the alcohols present in a wax[1].   It may also be employed to indicate the extent to which mono- and di-glycerides are present.   The hydrogen of each (OH) group is replaced by an acetyl radicle (.CH_3CO) on boiling with acetic anhydride.   Thus :

$$[R(OH)COO]_3C_3H_5 + 3 \genfrac{}{}{0pt}{}{CH_3CO\searrow}{CH_3CO\nearrow} O$$

1 *mol. of a glyceride*   3 *mols. of acetic*
of any hydroxylated   *anhydride*
acid

$$= [R(OCH_3CO)COO]_3C_3H_5 + 3CH_3COOH$$

1 *mol. of*   3 *mols.*
acetylated glyceride   *acetic acid*

On saponification, the acetyl group is split off:

$$[R(O.CH_3CO)COO]_3C_3H_5 + 6KOH$$

1 *mol. acetylated*   6 *mols. caustic*
glyceride   *potash*

$$= C_3H_5(OH)_3 + 3R(OH)COOK + 3CH_3COOK$$

1 *mol.*   3 *mols. potassium*   3 *mols. potass.*
glycerol   *soap of fatty acid*   *acetate*

---

[1] Also, of course, of hydroxylated fatty acids, if present.

On adding the equivalent amount of H_2SO_4:

$$H_2SO_4 + R(OH)\,COOK + CH_3COOK$$

1 *mol.*   1 *mol. soap*   1 *mol.*
sulphuric   *potass. acetate*
acid

$$= K_2SO_4 + CH_3COOH + R(OH)\,COOH$$

1 *mol.*   1 *mol. acetic*   1 *mol.*
potass.   *acid*   *fatty acid*
sulphate

The reaction with ALCOHOLS is as follows:

$$C_{16}H_{33}OH + (C_2H_3O)_2O = C_{16}H_{33}O\,(C_2H_3O) + CH_3COOH$$

1 *mol.*   1 *mol. acetic*   1 *mol. cetyl acetate*   1 *mol.*
cetyl alcohol   *anhydride*   *acetic acid*

$$C_3H_5(OH)_3 + 3\,(C_2H_3O)_2O = C_3H_5(O.C_2H_3O)_3 + 3\,CH_3COOH$$

1 *mol.*   3 *mols. acetic*   1 *mol. triacetin*   3 *mols.*
glycerol   *anhydride*   *acetic acid*

The latter reaction is made use of in the *acetin* method for glycerin analysis (page 188).

Oils which contain free fatty acids give variable acetyl values owing to the presence of mono- and di-glycerides. The reactions here are as follows:

$$C_3H_5\!\!\begin{array}{l}OH\\OH\\O.C_{18}H_{35}O\end{array} + 2\,(C_2H_3O)_2O$$

1 *mol.*   2 *mols. acetic*
monostearin   *anhydride*

$$= C_3H_5\!\!\begin{array}{l}O.C_2H_3O\\O.C_2H_3O\\O.C_{18}H_{35}O\end{array} + 2\,CH_3COOH$$

1 *mol. diacetyl*   2 *mols.*
monostearin   *acetic acid*

$$C_3H_5\!\!\begin{array}{l}OH\\O.C_{18}H_{35}O\\O.C_{18}H_{35}O\end{array} + (C_2H_3O)_2O$$

1 *mol. distearin*   1 *mol. acetic*
anhydride

$$= C_3H_5\!\!\begin{array}{l}O.C_2H_3O\\O.C_{18}H_{35}O\\O.C_{18}H_{35}O\end{array} + CH_3COOH$$

1 *mol. mono-*   1 *mol.*
acetyl distearin   *acetic acid*

Pure tri-glycerides[1], containing no hydroxyl groups, give a negative acetyl value, but the STEROLS present in oils and fats contain a hydroxyl group, and on this account all oils and fats give small positive values.

[1] Lewkowitsch has shown that the method of *Benedikt* and *Ulzer*, who acetylated the mixed fatty acids, yields unreliable results, owing to the formation of anhydrides of the fatty acids by the action of the acetic anhydride.

In the case of oils and fats containing VOLATILE FATTY ACIDS these are liberated on saponification, and thus increase the apparent acetyl value. A blank determination should therefore be made, and the figure obtained subtracted from the acetyl value found. This gives the TRUE ACETYL VALUE.

§ 4. **Apparatus and Materials required**

[A] Acetic anhydride.
> Pure and free from acetic acid.

[B] Alcoholic potash, standardised accurately.
> About normal is suitable.

[C] Sulphuric acid, N.
> Normal solution.

[D] Aqueous standard caustic potash, N/10.

[E] Round-bottomed flasks, 200 c.c. capacity.

[F] Reflux condenser.

[G] Beaker of 1 litre capacity.

[H] Evaporating basin (large).

[I] Kipp's CO_2 generating apparatus.

[J] A finely drawn out, bent tube.
> This is for passing CO_2 into solution in beaker.

[K] Funnels, filter papers, drying oven.

[L] Burettes and pipettes.

§ 5. **Procedure**

1. Take about 10 grms of the oil to be tested, and place in round-bottomed flask [E].

2. Add 20 c.c. of acetic anhydride [A], connect to reflux [F], and boil on sand-bath or over small naked flame for two hours.

3. Transfer to large beaker [G] and add 500 c.c. of boiling water.

4. Connect Kipp [I] with bent tube [J] and pass CO_2

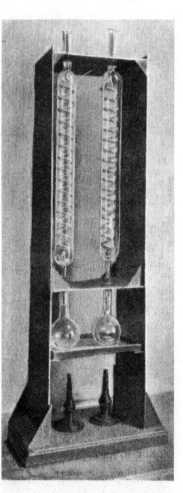

Fig. 48

through, while boiling, for half an hour (the object of this is to prevent bumping).   Then allow to stand for a few minutes.

5. Siphon off bottom aqueous liquor, and boil out again three successive times till the water is free from acetic acid (test with litmus).

6. Now filter the acetylated oil through a dry filter paper in drying oven.

7. Weigh out accurately about 5 grms of the acetylated product and add 50 c.c. of standard alcoholic potash [B].

8. Boil under reflux for one hour: then transfer to large evaporating basin, evaporate off alcohol, add water and dissolve soap.

9. Add the equivalent volume of N. H_2SO_4 corresponding to the potash used (7) plus 0·5 c.c. over.

10. Warm gently: the fatty acids will collect on top.

11. Filter through a wet filter paper into titrating flask, and wash fatty acids with hot water till the filtrate is no longer acid.

12. Titrate filtrate with $N/10$ KOH using phenol phthalein as indicator.

13. From result obtained subtract 5·0, being allowance for excess of N. H_2SO_4 added (9).

14. Acetyl value
$$= \frac{\text{No. of c.c. } N/10 \text{ KOH} \times 5\cdot61}{\text{Weight of acetylated oil taken}}.$$

If the oil contains volatile fatty acids, the original weight of the oil taken in (1) must be accurately weighed, and the weight of the yield of acetylated product obtained.

An equal weight of oil is then saponified and the estimation of the volatile fatty acids proceeded with, using directions 7 to 13 which apply equally.

This result must then be subtracted from the previous one in order to obtain the *true acetyl value*.

§ 6.   **Example**

Weight of acetylated fat + flask   27·320 grms,
Tare flask   22·297
Weight of acetylated fat taken   5·023

No. of c.c. of $N/10$ KOH used in titration   25·8 c.c.
less 5·0 c.c. for excess H_2SO_4 used   20·8

Acetyl value

$$\frac{20\cdot8 \times 5\cdot61}{5\cdot023} = \mathbf{23\cdot2}.$$

Mgs of KOH required to neutralise volatile
acids from 1 grm of fat   5·9
True acetyl value

$$23\cdot2 - 5\cdot9 = \mathbf{17\cdot3}.$$

Students should cleanse all apparatus in hot soft soap solution, and rinse in water.

THE ACID VALUE ⑭

§ 1. Definition

The acid value of an oil, fat, or wax is a measure of the free fatty acidity, and represents the number of milligrammes of potassium hydroxide required to neutralise the free fatty acids in 1 gramme of the substance.

§ 2. Outline of Method

A weighed quantity of the substance is treated with alcohol and warmed to effect solution of the free fatty acids.   It is then titrated with aqueous $N/10$ NaOH, using *phenol phthalein* as indicator.

§ 3. Remarks

Commercial samples of oils and fats usually contain more or less free fatty acids.   This is due to hydrolysis of the glycerides on contact with air, apparently by the action of minute quantities of enzymes derived from the tissues of the plants and animals.   In the case of vegetable oils, however, a slight degree of acidity is almost always met with in the freshly expressed oils.   The degree of acidity is greater in unripe seeds and fruits and diminishes as the fruit ripens.

Since *rancidity* is accompanied by the formation of free fatty acids— though the acidity does not of itself account for the phenomena[1] accompanying rancidity—the test often serves to determine the QUALITY of the fat and to define its commercial grade.   It should however be borne in mind that acidity in an oil or fat is often accompanied by

1.   More or less breaking down of the higher fatty acids into lower (volatile) acids.
2.   Partial splitting of the glycerides resulting in the formation of mono- and di-glycerides.
3.   Oxidation of the liberated fatty acids.
4.   Decomposition of the glycerol produced.

Thus oils and fats which have identical acid values may vary a good deal in quality, and the acid value is not, in itself, an absolute criterion of quality, though it gives, in most cases, a fair indication.

The WAXES (animal and vegetable) are usually characterised by a fairly constant percentage of free fatty acids, and they do not tend to hydrolysis on exposure or on keeping.

The *solvent* employed for the test—alcohol or methylated spirit— should be carefully neutralised, using $N/10$ alkali and phenol phthalein.

[1] See Vol. I, page 16.

For some hard waxes a mixture of benzene and absolute alcohol is necessary for clear solutions. The *standard alkali* may be aqueous or alcoholic, but the former is usually[1] preferable. If the acidity is small N/10 alkali should be employed; if large in amount, N/2 alkali is preferable. At the end of the titration there should be at least 50 PER CENT. OF ALCOHOL present. The end point is quite definite, but must be taken as the *first persistent pink coloration* obtained on shaking, as the coloration, especially in warm and strongly alcoholic solutions, rapidly disappears owing to saponification of the glycerides or esters.

Mineral acids should be tested for by shaking the melted fat with distilled water and testing the aqueous liquid, and if present, they may be estimated by titrating this with N/10 alkali, using *methyl orange* as indicator.

§ 4. Apparatus and Materials required

[A] Standard NaOH or KOH, N/10 and N/2 (aqueous or alcoholic, *v.s.*).

[B] Alcohol, or methylated spirit, accurately neutralised; for some waxes a mixture of absolute alcohol 2 to benzene 1.

[C] Phenol phthalein 1 per cent. in alcohol.

[D] Accurate burette and pipettes.

[E] Titrating flasks.

§ 5. Procedure

1. To a 250 c.c. titrating flask add about 5 grms of the oil or fat, accurately weighed.

    This, in the case of an *oil*, is best accomplished by weighing a small beaker containing about 10 grms of the sample, pouring out about half into the flask (*avoiding the loss of drops from beaker*) and reweighing. The difference = oil added. In case of a *fat*, a small evaporation basin may be employed with a spatula, weighing basin, fat and spatula before and after removal of the fat.

2. Add 50 c.c. of neutral alcohol or methylated spirit [B] and bring to a boil.

3. Add 1 c.c. of phenol phthalein [C] and titrate with N/10 or N/2 alkali. Add the alkali in drops at a time, well agitating flask during the addition. Note number of c.c. of alkali employed.

4. Acid value

$$= \frac{\text{No. of c.c. used} \times 5\cdot61}{\text{Weight of fat taken}} \quad \text{[with N/10 alkali]}$$

$$= \frac{\text{No. of c.c. used} \times 28\cdot05}{\text{Weight of fat taken}} \quad \text{[with N/2 alkali]}.$$

§ 6. Example

Weight of beaker + oil (i) = 25·073 grms,

„   „   „   (ii) = 20·015,

∴ weight of oil = **5·058**.

---

[1] Except in the case of high melting point waxes.

Volume of N/10 KOH used = 7·5 c.c.

Now 1 c.c. of N/10 KOH contains ·0056 grm KOH.

∴ 7·5 c.c. KOH contain 7·5 × 5·6 mg. KOH.

Hence 5·058 grms of oil require 7·5 × 5·6 mg. KOH.

∴ 1 grm of oil requires $\dfrac{7·5 \times 5·6}{5·058}$ mg. KOH.

∴ acid value = **8·30**.

§ 7. Expression of Results

Acidity is often expressed in different terms in practice, and the commonest method is to give the figure in terms of *percentage of oleic acid*.

In this case the calculation is made as follows :

$$\frac{\text{Number of c.c. of N/10 alkali} \times 2·82}{\text{Weight of fat taken}}$$

(since 282 is the molecular weight of oleic acid).

Or the acid value may be multiplied by the factor 0·5027.

Students should cleanse all apparatus in hot soft soap solution, and rinse in water.

UNSAPONIFIABLE MATTER (15)

§ 1. Definition

The "UNSAPONIFIABLE MATTER" in an oil, fat or wax is usually held to comprise those substances which are not water-soluble after treatment with alkalies (saponification). It is expressed in percentage of the oil, fat, etc.

§ 2. Outline of Method

A weighed quantity of the substance is saponified with alcoholic potash, the alcohol evaporated off, and the soap paste either mixed with sand and treated in an extractor with ether, or the soap dissolved in water, and the solution shaken in a separating funnel with successive small quantities of solvent. In each case the solvent, on evaporation, yields the "unsaponifiable matter" in the substance.

§ 3. Remarks

As previously stated (Vol. I, pages 15, 62) all animal oils contain small quantities of CHOLESTEROL or related alcohols, and all vegetable oils up to two per cent. of PHYTOSTEROLS.

These substances form usually the greater part of the unsaponifiable matter of genuine oils and fats.

Spermaceti is present in some marine oils, and shark liver oil of certain species consists principally of the unsaturated hydrocarbon *Spinacene*.

In addition, *resinous substances* of an undetermined character often accompany the sterols and render the separation of these difficult. *Albuminous impurities* derived from the seeds of the plant or from the tissues of the animal usually appear, in the determination of this value, as a flocculent middle layer between the aqueous soap solution and the ether ; it is not usual to include this in the amount of "unsaponifiable matter." If required it may be estimated by filtration of the aqueous portion after separation from the ether.

The NATURAL WAXES yield high proportions of unsaponifiable matter, due to the insoluble alcohols contained free in the wax, or liberated by saponification. Varying proportions of hydrocarbons are also frequently present.

The MINERAL OILS AND WAXES in the refined state contain only minute amounts of saponifiable constituents. The less refined oils contain varying amounts of bituminous matters which yield soluble compounds with potash, and unrefined coal-tar products contain proportions of phenols and acid substances which are also removed with alkalies.

Brown coal, and other BITUMINOUS WAXES are largely saponifiable, being esters of montanic acid, and the refined products of these usually contain the free acid and hydrocarbons. The separation of the latter may be effected in this manner.

Rosin oils, although mainly unsaponifiable, contain varying proportions of rosin (abietic) acid which combines with potash in this test.

An unduly high figure in the case of the animal and vegetable oils indicates adulteration with mineral or rosin oils, or possibly with shark liver oil.

In order to confirm this, it is necessary to further examine the unsaponifiable matter (see page 258).

As regards the *method* employed, the authors prefer, in most cases, the wet extraction, as being quicker and generally more reliable (as here described), providing the modification suggested by *Wilkie*[1] be employed, the formation of troublesome emulsions being thus avoided.

The extractor recommended for the dry method is one of those described on pages 21 and 22. In the case of *waxes* the dry method should be employed, and the *Stokes'* extractor (in which the thimble containing the material is actually immersed in the boiling solvent) is most suitable.

Lewkowitsch[2] recommends, in the case of waxes, to obtain the calcium or barium salts by precipitation of the neutralised aqueous soap solution with $CaCl_2$ or $BaCl_2$. The precipitate is washed, dried, mixed with sand and boiled out (in Stokes' extractor) with petroleum ether.

The method of *Wilkie* may be used, which consists in mixing the wax with oil before saponification and wet extraction.

The most suitable *solvent* to employ is probably *petroleum ether* fractionated to distil under 40° C. This dissolves much less soap than does methylated ether. The latter however is always safer, as it dissolves a wider range of products insoluble in water (see definition, § 1) than petrol. ether. *Lewkowitsch*[3] recommends to incinerate the unsaponifiable matter obtained, lixiviate the ash with hot water, filter, and titrate the solution with N/10 acid, calculating the alkali found to soap and deducting this figure from the weight of the unsaponifiable matter.

§ 4. Apparatus and Reagents required

[A] 200 c.c. flask.

[B] Reflux condenser.

[C] Separator.

[D] Large porcelain basin.

[E] Fat extractor (*Bolton* and *Revis* or *Soxhlet*). (See pages 21 and 22.)

[F] Alcoholic potash 2N.

> Place several sticks of potash in a large beaker, cover with absolute alcohol, protect from evaporation by a large clock glass, and allow to stand till saturated. Titrate with N. HCl, and adjust to 2N with absolute alcohol.

---

[1] *Analyst*, 1917, **42**, 200.
[2] Lewkowitsch, *Oils, Fats and Waxes*, Vol. I, p. 400.   [3] *Ibid.* p. 457.

[G] Ether.

[H] Alcohol 95 per cent., or purified methylated spirit.

[I] Sodium bicarbonate.

[J] Petroleum ether fractionated, using a good column, and rejecting
fractions boiling over 40° C.

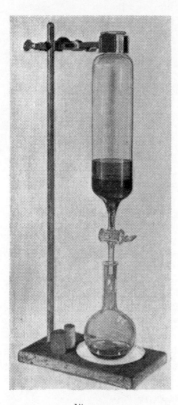

Fig. 49

§ 5. **Procedure**

A. WET EXTRACTION

1. Tare the 200 c.c. flask, and weigh into it accurately about 5 grms of the oil (*a*).

2. Add 12·5 c.c. 2N alcoholic potash [F], and connect up to reflux condenser. Boil gently (water- or sand-bath) for $1\frac{1}{2}$ hours.

3. Transfer the alcoholic soap solution formed to a separator with 50 c.c. of water; cool. Add 40 c.c. ether, shake, allow to separate and run off soap liquor into flask, and ether layer into a stoppered bottle.

4. Replace soap liquor in separator and extract again with 30 c.c. ether[1]. Run off as before, add ether extract to previous charge.

5. Treat soap liquor once more with 30 c.c. ether, unite ether extracts, and pour into a separator containing 20 c.c. of distilled water. Run off water without shaking.

6. Wash the ethereal solution by shaking vigorously with the following quantities of distilled water, allowing a clean separation each time and then running away the water layer.

> 2 c.c. distilled water,
> 5 c.c.   „   „
> 30 c.c.   „   „

[1] If the separation of the ether layer is at all sluggish, heat soap liquor on water-bath till the dissolved ether commences to boil, and cool again.

7. Run the ether into a tared flask, distil off by attaching to a condenser and applying hot water to the flask, and dry the residual matter in a water oven to constant weight (*b*).

$$\frac{b \times 100}{a} = unsaponifiable\ per\ cent.$$

Notes.—As Wilkie[1] has shown, the quick separations depend on the proportion of alcohol present in the soap liquor. The quantities stated must therefore be strictly adhered to.

In case of *waxes*, Wilkie advises to use a mixture of, say, 0·5 grm wax with 4·5 grms castor oil, and saponify with 12·5 c.c. 2N alcoholic potash (as above), but to use 40 c.c. of water at 30° C. instead of 50 c.c. cold water, and extract with successive volumes of 50, 40, 40, 30 c.c. ether. Wash with 2, 5, 30 c.c. water, evaporate and weigh. A correction is applied for the known unsaponifiable matter of castor oil.

B. DRY EXTRACTION

1. Place about 10 c.c. of the oil in a weighing bottle ; weigh accurately, pour out about half into a large porcelain basin and weigh again. Difference (*a*).
2. Dissolve 2 grms of caustic soda in 2 c.c. water, add to oil, with 100 c.c. of 95 per cent. alcohol. Place an inverted funnel over basin, and heat on a water-bath, stirring occasionally with a glass rod, for ¾ hour.
3. Evaporate to a pasty consistency ; then add about 10 grms sand and 2—3 grms sodium bicarbonate, mix well and heat to dryness over the water-bath.
4. Grind up the lumps to powder, dry well in a water oven, and place in a fat extractor; extract 2—3 hours with petroleum ether (fractionated below 40° C. [J]) using a tared flask for the solvent.
5. Remove the material, dry by setting aside in the open for a few minutes, regrind, dry in water oven again and extract for a further hour.
6. Evaporate off ether in flask and dry in water oven to constant weight = unsaponifiable matter (*b*).

$$\frac{b \times 100}{a} = unsaponifiable\ per\ cent.$$

For the further examination of the unsaponifiable matter see page 176.

Students should leave flasks and apparatus clean, rinsing out with hot soap solution. Recovered ether should be returned to reagent bottle.

[1] *Analyst*, 1917, **42**, 200.

SECTION IV

SPECIFIC TESTS FOR OILS, FATS AND WAXES

SECTION IV

SPECIFIC TESTS FOR OILS, FATS AND WAXES

CLASS I. MARINE OILS

GROUP I. FISH OILS

Menhaden Oil

HOPPENSTEDT'S COLOUR REACTION

Procedure

1. Place in a test-tube 5 c.c. of oil, 5 c.c. acetone, and 1 c.c. of conc. hydrochloric acid, and shake for 1 minute.
2. Add to the tube 5 c.c. of petroleum ether, and mix by inverting tube twice: allow to settle.
3. The bottom layer, in the case of menhaden oil, is coloured a *bluish-green*, while with cod liver oil, it is brown or yellow.

Remarks

In the case of mixtures containing less than 40 per cent. of menhaden oil, the green colour becomes masked by the brown tint of the oil.

GROUP II. LIVER OILS

SULPHURIC ACID TEST FOR LIVER OILS

Procedure

1. Dissolve about 2 c.c. of the oil in about 10 c.c. of carbon disulphide*, and pour a little conc. sulphuric acid slowly down the side of the tube.
2. A *blue* coloration at the junction of the acid and oil indicates a liver oil.

Remarks

In the case of very rancid oils, a purplish or brown coloration is obtained.   Some blubber oils[1] are said to give the blue colour.

---

* *Caution, highly inflammable.*
[1] Thompson and Dunlop.

CLASS II.   DRYING OILS

Chinese Wood (Tung) Oil

SOLIDIFICATION TEST

Procedure
1. Dissolve about 2 c.c. of the oil in 5 c.c. chloroform in a small basin, and add 5 c.c. of a saturated solution of iodine in chloroform. Stir with a glass rod.
2. In about 2 minutes a *stiff jelly* is produced with tung oil.

Walnut Oil

Detection of Poppy seed oil in Walnut oil (Bellier[1])

Procedure
1. Dissolve 16 grms of KOH in 100 c.c. of 92 per cent. alcohol.
2. Warm 1 c.c. of the oil with 5 c.c. of (1) until a clear solution is obtained, when a cork is inserted and it is further heated to ·70° C. for half an hour in a water-bath.
3. Sufficient of a 25 per cent. acetic acid solution is now added to exactly neutralise the 5 c.c. of alcoholic potash (ascertained by a blank test).
4. The test-tube is again corked, and placed in water of 25° C., then in water of 17°—19° C., shaking frequently.
5. In the case of pure walnut oil, only a very small amount of precipitate is formed ; poppy seed oil gives a *copious ppt.*

Remarks
Not less than 20 per cent. of poppy seed oil can be detected by this means[2].

CLASS III.   SEMI-DRYING OILS

Cottonseed Oil

HALPHEN COLOUR TEST

Procedure
1. Make a 1 per cent. solution of flowers of sulphur in carbon disulphide*.

---

[1] *Ann. Chim. Appliq.* 1905, 52.
[2] Balavoine, *Jour. Suisse de Chimie et Pharmacie*, 1906, 15.
* *Caution, highly inflammable.*

2. Take a small quantity of oil, dissolve in an equal volume of amyl alcohol, and add the same volume of (1).
3. Immerse test-tube in boiling water.
4. Cottonseed oil gives, in from 5 to 30 minutes, a *deep red coloration* (sometimes orange-red).

Remarks

The amyl alcohol may be replaced by a drop of pyridine.

Five per cent., or less, of cottonseed oil in another oil may be detected by this reaction.

The colour-producing substance passes into the fat of hogs etc. fed on cotton cake, the lard from which gives the reaction. The reaction is rendered negative by heating the oil to 250° for 10 minutes, or by blowing the oil.

KAPOK oil also gives the red coloration.

NITRIC ACID TEST

Procedure

1. Into a test-tube, place a few c.c. of the oil, and an equal volume of nitric acid (preferably 1·37—1·41 S.G.), and shake thoroughly for at least two minutes. Allow to stand.
2. A *coffee brown* coloration indicates cottonseed oil.

Remarks

The test is said to detect 10 per cent. of cottonseed oil in olive oil. American cotton oil is less responsive to the test, and heating the oil renders it negative. Rape oil is also said to give the same coloration.

SILVER NITRATE (Becchi-Milliau) TEST

Procedure

1. The mixed fatty acids from the oil are obtained, and 5 c.c. dissolved in a test-tube in 15 c.c. of 90 per cent. alcohol. 2 c.c. of 3 per cent. silver nitrate solution are added.
2. The mixture is heated on the water-bath till about one-third of the alcohol is evaporated.
3. In the presence of cottonseed oil, darkening occurs, and the fatty acids, as they separate, are coloured *brownish* or black.

Remarks

Becchi originally used the *oil* for testing, and employed a reagent containing (in addition to nitric acid) amyl alcohol, ether, and colza oil.

Holde states that the test recognises as little as 1 per cent. of cottonseed oil in a mixture. Heated oils fail to give the reaction.

Sesamé Oil

BAUDOUIN TEST[1]

Procedure

1. Prepare a 2 per cent. solution of FURFUROL in alcohol.
2. Place in a test-tube 5 c.c. of oil, 5 c.c. of hydrochloric acid (S.G. 1·19) and 8 drops of (1) solution. Shake vigorously for half a minute and allow to settle.
3. In the presence of sesamé oil, the acid layer becomes a distinct *carmine red* colour.

Remarks

The reagent may also (as was originally proposed) be prepared by dissolving 0·1 grm of cane sugar in 10 c.c. of hydrochloric acid of S.G. 1·19. The use of furfurol is however preferable. The test will recognise as little as 0·5 per cent. of sesamé in other oils.

Butterfat may give the reaction if the cows have been fed with sesamé cakes.

The Soltsien Reaction

In cases where the fat has been coloured by the use of an aniline dye, which with hydrochloric acid gives a pink coloration, it is advised to use the following test, due to *Soltsien*[2].

A small quantity of oil or fat is dissolved in twice its volume of petroleum ether. It is then shaken with half this quantity of a concentrated solution of stannous chloride, saturated with HCl gas, heated to 40° till separation occurs and then immersed in water of 80° up to the limit of the aqueous layer. Sesamé oil gives a *red coloration* of the stannous chloride solution.

CLASS IV

Group I

Rape Oil

1. TORTELLI AND FORTINI TEST

Remarks

The detection and determination of rape oil in other oils are based upon the recognition of the characteristic acid, *Erucic Acid*, which exists as a glyceride in this oil. As erucic acid is a "solid acid" (M.P. 33°—34° C.), and also unsaturated (absorbs Br_2), its lead salt is partially soluble in cold ether, as distinct from the liquid unsaturated acids—the lead salts of which

---

[1] Villavecchia and Fabris' modified test, *Ztsch. f. angew. Chem.* 1893, **6**, 505.
[2] See Beythien, *Chem. Ztg.* 1909, **24**, 1019.

are entirely soluble—and from the solid saturated acids—which give, with lead, salts insoluble in ether.

It follows that the "solid acids" (lead salts insoluble) of an oil will, in the presence of rape oil, show an abnormally high Iodine Value, whilst the liquid acids will contain more "solid acids" than normal. *Tortelli* and *Fortini* ascertain the "turbidity temperature" of the alcoholic solution of the sodium salts of the liquid fatty acids as an indication of the presence of "solid" acid.

The authors have examined the original method, and find it somewhat lengthy and cumbersome. They have therefore modified it in several respects (as indicated).

Outline of Method

The lead salts, obtained by treatment of the saponified oil with lead acetate, are resolved by means of ether into soluble and insoluble portions. These are decomposed with hydrochloric acid, the M.P. and the iodine value of the insoluble fraction obtained, and the turbidity temperature of the alcoholic solution of the sodium salts of the soluble acids ascertained.

Apparatus and Reagents required

[A] Reflux condenser, fitted with 200 c.c. flask. (See page 106.)

[B] Conical, wide-necked flask, 500 c.c. capacity.

[C] Funnel, with tight cover (watch glass).

[D] Separator, 500 c.c. capacity.

[E] 2N Alcoholic Potash (90 per cent. alcohol).

> Dissolve 56·0 grms of KOH ("pure from alcohol") in 45 c.c. water; make up to 500 c.c. with absolute alcohol, *or* dissolve the potash in its own weight of water, and make up to 500 c.c. with purified "industrial" methylated spirit (non-mineralised) about 91 per cent. strength.

[F] Sodium Alcoholate, prepared by dissolving carefully ·6 grm clean metallic sodium in 100 c.c. of absolute alcohol. It must be freshly prepared.

[G] Absolute Alcohol.

[H] Acetic Acid, 10 per cent. solution.

[I] Ether (dry), methylated.

[J] Lead Acetate, pure crystalline.

[K] Phenol Phthalein, 1 per cent. in *absolute alcohol*.

[L] Hydrochloric Acid, 10 per cent. and 20 per cent. solutions.

Procedure

1. Into the 200 c.c. flask weigh 20 grms of the oil, add 50 c.c. of 2N alcoholic potash [E] and saponify under reflux. (See page 106.)

2. Add a few drops of phenol phthalein solution, and neutralise by titrating with 10 per cent. acetic acid [H].

3. Into flask [B] place 20 grms lead acetate, and dissolve in 300 c.c. of boiling distilled water. Slowly pour the neutralised soap liquid

into the boiling acetate, shaking constantly as the lead salts separate.

4. Cool under the tap, rotating the flask meanwhile to cause the salts to adhere to the sides of the flask.

5. Pour off the aqueous liquor into a large beaker, returning any particles of lead salts that run off to the flask. Wash the lead salts with 200 c.c. of warm water (70° C.) three times, finally draining flask, and removing any adherent moisture from the lead salts by means of a wad of dry cotton wool.

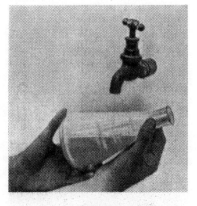

Fig. 50

6. Add to the flask 80 c.c. of ether, cork securely and well shake till the salts are broken up. Place under reflux condenser and heat by applications of hot water to flask, shaking at intervals, for 15 minutes.

7. Cork flask, and place in water at 15° C. for one hour, then pour contents on to a filter in a funnel placed in the mouth of separator [D] and cover with watch glass till ethereal solution has filtered through.

8. The filter paper with the insoluble lead salts is transferred to the flask again, boiled and cooled as before with a further 40 c.c. of ether, and filtered through a fresh paper. After draining, any remaining lead salt in the flask is washed out on to filter with a further 40 c.c. of ether, and this is allowed to drain through.

A. Insoluble Salts

9. The filter paper containing the insoluble lead salts is placed in a beaker, a little pumice powder added, and boiled for half an hour with 50 c.c. of 20 per cent. HCl solution, making up for any loss by evaporation of HCl.

10. The liquid is run into a separating funnel, cooled to 15°, and 50 c.c. of ether is then added, the separator shaken and allowed to stand. The aqueous liquor and precipitated lead chloride is then run off, the ether layer shaken with 20 c.c. of 10 per cent. HCl, settled, the acid removed, and the ether washed with three successive lots of 20 c.c. water.

11. The ethereal solution is run into a small flask, the solvent evaporated off, and the acids dried at 100° C. in the water oven.

12. The M.P. of the fatty acids and the Iodine Value are then determined.

Tortelli and Fortini obtained the following results:

Oil	Iodine value	Melting point °C.
Olive	7·8	58—59
Rape	62·0	41—42
50 per cent. Rape in Olive	32·0	47—48
30 ,, ,, ,, ...	28·0	48—49
20 ,, ,, ,, ...	22·1	50—51
10 ,, ,, ,, ...	12·8	54—55
Sesamé	9·3	55—56
Arachis	13·0	57—58
Cotton	19·0	57—58

B. Soluble Salts

13. To the ethereal filtrate in the separator (see 8 above) is now added 150 c.c. of 10 per cent. hydrochloric acid, the stopper inserted and the whole well shaken. The fatty acid solution is allowed to separate from the milky lead chloride and water, and the latter then run off. A further 150 c.c. of acid is then added, the separator again well shaken, and the liquor run off.

14. The ethereal solution is washed with two lots of 100 c.c. water, and is then run into a dry tared flask, the ether evaporated by immersion of the flask in warm water, and the acids dried.

If the oil under examination is a drying or semi-drying oil, this is best done in a current of CO_2 or hydrogen (see page 18). If a non-drying oil, it may be safely carried out in a water oven at 100° C.

15. Into a tared 50 c.c. measuring flask is now accurately weighed out 1·166 grms[1] of the dry acids, 10 c.c.[2] of absolute alcohol added, also a few drops of phenol phthalein in absolute alcohol [K], and the acid then carefully neutralised with the sodium alcoholate [F], the solution being meanwhile kept clear by warming. Absolute alcohol is now added up to the 50 c.c. mark, the solution shaken, a suitable quantity run into a test-tube, and the turbidity temperature ascertained as in the *Valenta test* (page 80).

Fig. 51

[1] Tortelli and Fortini obtain the sodium salts and dry these over sulphuric acid *in vacuo*. The authors find that this is unnecessary, but in order to conform to the original quantities of sodium salts given, they have calculated the weight of the fatty acids corresponding approximately to the original directions.

[2] It may be more convenient to use the whole bulk of the dry acids, already

Tortelli and Fortini obtained the following results:

Oil					Turbidity Temp.
Olive ...	...	...	...	...	20—24°
Rape ...	...	...	...	...	45—50°
50 per cent. Rape in Olive		...	...	...	35—40°
20 ,, ,, ,,			...	...	30—35°
10 ,, ,, ,,			...	...	30—34°
Cotton ...	...	...	...	...	14—16°
Sesamé ...	...	...	...	...	18—20°
Arachis ...	...	...	...	...	18—22°

NOTE. Since (like the *Reichert-Meissl*) this test depends upon the accurate fulfilment of the conditions imposed, it is advisable to check results with a sample of genuine rape oil.

2. Holde and Marcusson's Test[1]

Remarks

This method is based upon the fact that erucic acid is more soluble in chilled 96 per cent. alcohol than are the saturated solid acids.

Procedure

1. The mixed fatty acids are obtained from the oil by saponification and acidification of the aqueous soap solution with a mineral acid in the usual way.

2. 25 grms of the mixed fatty acids are dissolved in 50 c.c. of 95 per cent. (by volume) alcohol and cooled with stirring to – 20° in an ice and salt freezing mixture.

3. The precipitate is filtered through a funnel placed in a freezing mixture and connected with a suction flask, and is washed with chilled alcohol.

4. The filtrate is evaporated, the residue dissolved in 75 per cent. alcohol, and cooled to – 20°. In the presence of 20 per cent. of rape oil a crystalline precipitate separates in the course of an hour.

5. The ppt. is filtered as before, washed with chilled alcohol (75 °/₀), dissolved in warm ether, and the solvent evaporated. It is weighed, dissolved in alcohol, titrated with N/10 alcoholic potash, and the molecular weight determined.

NOTE. If the oil is rich in solid fatty acids (e.g. whale oil) on cooling to – 20° the ppt. will be so bulky as to render filtration difficult. In this case, a preliminary cooling to 0° and filtration is advisable, the filtrate being afterwards treated as above.

Holde obtained from a specimen of rape oil 0·74 grm of crude erucic acid of M.P. 30° and mol. wt. 321·5.

weighed, calculating the amount of alcohol necessary (= 42·9 c.c. for each grm of fatty acids), adding half this before neutralising and the rest after, less the number of c.c. of alcoholate used. This method gives slightly more alcohol than above, but this does not affect the turbidity figure.

[1] *Zeit. f. angew. Chem.* 1910, **23**, 1260.

GROUP II

Kernel Oils

BIEBER'S TEST

Procedure

1. Prepare the reagent by weighing in a beaker equal weights of water, conc. sulphuric acid, and fuming nitric acid, in the order named.

   The weighing should be done on a rough balance, and the acids added drop by drop, stirring with a glass rod.

2. Place 1 c.c. of oil in a test-tube, and add 2 c.c. of reagent from a graduated cylinder. Shake well, and allow to stand.

3. The kernel oils give, after standing, a *faint pink* coloration.

Remarks

Lewkowitsch[1] could not detect less than 25 per cent. of apricot kernel oil in almond oil with the test.

The reagent should be prepared afresh each time.

Arachis Oil

The special tests for the recognition and determination of Arachis Oil depend upon its content of glycerides of Arachidic and Lignoceric acids, and the relative insolubility of these in solvents.

BELLIER'S TEST

This test, originally proposed by *Bellier*, has been modified in several respects by other observers[2], and has been examined by Evers[3] who found the modification here described satisfactory. The authors have also confirmed this in their own experience.

Remarks

The exact procedure must be adhered to in order to obtain reliable results. The test is not suitable for detection of arachis oil in fats, with the exception of palm kernel and coconut, but is suitable for all oils.

Reagents required

[A] Alcoholic Potash.

   Dissolve 80 grms of KOH, pure from alcohol, in 80 c.c. water, and make up to 1 litre with 90 per cent. (by volume) alcohol.

[B] Dilute Acetic Acid.

   Dissolve 1 volume glacial acetic acid in 2 vols. water.

---

[1] *Analyst*, 1904, **29**, 106.

[2] Mansfeld, *Zeit. f. Unters. d. Nahrgs- u. Genussm.* 1905, **17**, 57; Alder, *ibid.* 1912, **24**, 676; Franz, *Beiträge z. Nachweis u. z. Kenntniss d. Erdnussöles*, Dissert., München, 1910.

[3] *Analyst*, 1912, **37**, 487.

[C] Alcohol, 70 per cent. (by volume).

> Place in a graduated 100 c.c. flask 30 c.c. water, and make up to mark with pure absolute alcohol [s.g. = ·8900].

Procedure (Evers' modified)

1. Measure accurately 1 c.c. of the oil, and place in a 100 c.c. flask, add 5 c.c. of alcoholic potash [A], insert cork with a long condensing tube, and saponify by heating for exactly 4 minutes over a boiling-water bath.

2. Cool the solution to 15° C. and add exactly 1·5 c.c. of dilute acetic acid [B] with 50 c.c. of 70 per cent. alcohol [C], and shake the liquid till it becomes clear.

   > If solution remains turbid, warm till it clears.

3. Immerse in cold water, shaking continuously, and NOTE THE TEMPERATURE at which turbidity occurs.

   > If fairly high, re-warm till clear, and repeat without cooling in water.

4. If turbidity does not appear at 16°, shake at this temperature for 5 minutes, and then lower to 15·5° C. In all cases where turbidity occurs, warm till clear, and repeat, taking as final the *second* turbidity temperature.

   > Some olive oils have been found to give a turbidity at first, but this is not obtained on repeating after re-warming. Others give a slight opalescence, which is not followed by a *distinct precipitate*, as in the presence of arachis oil.

5. Distinct turbidity is produced in the presence of 5 per cent. and upwards of arachis oil in the sample.

Approximate percentage of Arachis Oil present corresponding to temperatures of turbidity (Evers).

Oil		Turbidity Temp.	Oil		Turbidity Temp.
Olive ...	...	11·8—14·3	+ Arachis 50 °/₀		33·8
+ Arachis	5 °/₀	15·9—17·0	,,	60	35·3
,,	10	19·8	.,	70	36·6
,,	20	25·7	,,	80	38·0
,,	30	29·2	,,	90	39·3
,,	40	31·5	Arachis	...	40·0—40·8

EVERS' MODIFIED QUANTITATIVE METHOD

Additional Reagents

[D] Acid Alcohol.

> 70 per cent. (by volume) alcohol (C above) with 1 per cent. (vol.) Hydrochloric Acid. Place 1 c.c. HCl in 100 c.c. graduated flask, and fill up to mark with 70 per cent. alcohol.

[E] Alcohol, 90 per cent.

Procedure

1. Weigh 5 grms of the oil into a 100 c.c. flask, add 25 c.c. of alcoholic potash [A] and saponify by heating under a reflux condenser for 5 minutes.

2. To the hot soap solution add 7·5 c.c. dilute acetic acid [B] and 100 c.c. of acid alcohol [D] and cool to 12°—14° C. for 1 hour.

3. Filter, and wash with acid alcohol at 17°—19°, the ppt. being kept loose by means of a loop of platinum wire until the filtrate gives no turbidity with water. MEASURE THE WASHINGS.

4. Dissolve ppt. according to bulk in 25—70 c.c. hot 90 per cent. alcohol and cool to a fixed temperature between 15° and 20° C. If crystals appear in any quantity, allow to stand at this temperature 1 to 3 hours. Filter, wash with a measured volume of 90 per cent. alcohol (using about half the volume taken for crystallisation), and finally with 80 c.c. of 70 per cent. alcohol.

5. Wash crystals with warm ether into weighed flask. Distil off ether, dry at 100° C., and weigh.

6. Ascertain M.P. of crystals (page 39). If lower than 71° C., recrystallise from 90 per cent. alcohol. Add correction for solubility in 90 per cent. alcohol, as below (Table I), and for total volume of 70 per cent. alcohol used in precipitating and washing (Table II).

TABLE I. *Correction per* 100 *c.c. of* 90 *per cent. alcohol used for crystallisation and washing* (*Archbutt*).

Weights of fatty acids	Grms at		
	15° C.	17·5° C.	20° C.
0·1 or less	+0·033	+0·039	+0·046
0·2 ,, ...	0·048	0·056	0·064
0·3 ,, ...	0·055	0·064	0·074
0·4 ,, ...	0·061	0·070	0·080
0·5 ,, ...	0·064	0·075	0·085
0·6 ,, ...	0·067	0·077	0·088
0·7 ,, ...	0·069	0·079	0·090
0·8 ,, ...	0·070	0·080	0·091
0·9 and upwards	0·071	0·081	0·091

TABLE II. *Correction per* 100 *c.c. of* 70 *per cent. alcohol used for washing.*

Weights of fatty acids	Grms		
	M.P. 71° C.	M.P. 72° C.	M.P. 73° C.
Above 0·10 grm	0·013	0·008	0·006
0·08—0·10	0·011	0·007	0·006
0·05—0·08	0·009	0·007	0·005
0·02—0·05	0·007	0·006	0·005
Less than 0·02	0·006	0·005	0·004
Factor for conversion of percentage of fatty acids to Arachis Oil	17	20	22

Typical results—Evers' modification of Bellier's Test

Oil	Alcohol used for crystallisation Per cent.	Weight of Crystals	Correction for 90 per cent. Alcohol	Correction for 70 per cent. Alcohol	Total	Per cent.	Melting point C.	Per cent. of Arachis Oil by Factor
Arachis (A)	90	0·160	0·040	0·027	0·227	4·54	73	100
	70	0·218	—	0·065	0·283	5·66	71	96
Arachis (B)	90	0·163	0·045	0·032	0·240	4·80	72	96
	70	0·233	—	0·068	0·301	6·02	71	102
Arachis (C)	90	0·152	0·054	0·034	0·240	4·80	72	96
Arachis (D)	90	0·194	0·033	0·028	0·255	5·10	72	102
Arachis (A) 50 % / Olive "Nice" 50 %	90	0·056	0·032	0·022	0·110	2·20	73	48
	70	0·090	—	0·055	0·145	2·90	71	49
Arachis (A) 35 % / Olive "Nice" 65 %	90	0·045	0·020	0·029	0·094	1·88	71	32
	70	0·029	0·040	0·020	0·089	1·78	72·5	37
Arachis (A) 20 % / Olive "Nice" 80 %	90	0·059	—	0·040	0·099	1·98	71	34
	70	0·024	0·012	0·019	0·055	1·10	71	19
Arachis (C) 20 % / Olive "Malaga" 80 %	90	0·030	—	0·024	0·054	1·08	71	18
	70	0·012	0·020	0·015	0·047	0·94	72	19
Arachis (A) 10 % / Olive "Nice" 90 %	90	0·021	—	0·027	0·048	0·96	71	16
	70	0·009	0·008	0·008	0·025	0·50	73	11
Arachis (B) 10 % / Olive "Nice" 90 %	90	0·008	—	0·015	0·023	0·46	70	8
	70	0·012	—	0·018	0·030	0·60	71	10
Arachis (C) 10 % / Olive "Malaga" 90 %	90	0·011	—	0·016	0·027	0·54	71	9
	70	0·007	—	0·012	0·019	0·38	71	6·5
Arachis (A) 5 % / Olive "Nice" 95 %	90	0	—	—	—	0	—	—
	70	0	—	—	—	0	—	—
Sesamé	90	0	—	—	—	0	—	—
	70	0·012	—	—	—	0·24	64	—
Cottonseed	90	0·006	—	—	—	0·12	50—55	—
	70	0·014	—	—	—	0·28	64—67	—
Olive "saponified"	90	0·021	—	—	—	0·42	64—68	—
	70	—	—	—	—	—	—	—

7. In case there are no crystals from 90 per cent. alcohol, or only a very small amount, reduce strength of alcohol to 70 per cent. (31 c.c. water to 100 c.c. 90 per cent. alcohol). Crystallise at 17°—19° for 1 hour, filter, wash with 70 per cent. alcohol and weigh as before, adding correction for 70 per cent. alcohol (Table II). If the M.P. of acids is below 71° C., recrystallise from a small quantity of 90 per cent. alcohol or again from 70 per cent. alcohol.

The results obtained are shown in the table on previous page.

Example

Weight of fatty acids obtained ... ... 0·0355 grm.
M.P. of crystals ... ... 72° C.
Total volume of 70 per cent. alcohol used (including original 100 c.c.)

$$100 + 75 + 80 = 255 \text{ c.c.}$$

$$\text{Correction} = 255 \times \frac{·006}{100} = 0·0135 \text{ grm.} \quad \text{(Table II.)}$$

Total volume of 90 per cent. alcohol used

$$65 + 30 = 95 \text{ c.c.}$$

Temperature of alcohol = 15° C.

$$\text{Correction} = 95 \times \frac{·033}{190} = ·0313 \text{ grm.} \quad \text{(Table I.)}$$

Weight of fatty acids and corrections ... ... 0·0803 grm.
Factor (Table II) = 20.

∴ **Arachis oil per cent.** $= \frac{100}{5} \times ·0803 \times 20 =$ **32 per cent.**

RENARD'S METHOD

Definition

This method gives an approximate estimation of the Arachidic and Lignoceric acids present in an oil.

Outline of Method

The oil is saponified, the fatty acids liberated by hydrochloric acid, dissolved in ether, the ethereal solution evaporated, and the fatty acids, dissolved in 90 per cent. alcohol, precipitated with a solution of acetate of lead, the lead salts exhausted with ether, and the insoluble residue decomposed with hydrochloric acid under ether. The latter is distilled off, and the dry fatty acids dissolved in a measured volume of 90 per cent. alcohol, cooled, filtered, and washed with 90 per cent. and 70 per cent. alcohols, dried and weighed. Allowance is made for the acid dissolved by the washings, and the final weight, multiplied by a given factor, gives the percentage of arachis oil present.

Remarks

As for Bellier's test (page 140).

Reagents and Apparatus required

[A] N. Alcoholic Potash.

[B] Hydrochloric Acid, s.g. 1·10=[HCl 1·16, 60 per cent., Water 40 per cent.].

[C] Ether (methylated) dry.

[D] Lead Acetate, 20 per cent. solution.

[E] Alcohol, 90 per cent. by volume (s.g. ·8340).

[F] „ 70 „ „ (s.g. ·8900).

[G] Reflux condenser, Soxhlet extractor, etc.

Procedure

1. Saponify 10 grms of the oil (as in sapon. value, p. 106) with 75 c.c. of N. alcoholic potash; evaporate off alcohol; dissolve soap in water, rinse into a separator with hot water, and add excess of HCl.

2. Cool well (to 15° C.), shake with three lots of 20 c.c. ether, and wash ethereal solution with lots of 10 c.c. water till free from mineral acid.

3. Run the ether into a flask, distil off, and dry the fatty acids in water oven.

4. Dissolve dry fatty acids in 50 c.c. of 90 per cent. alcohol [E], cool to 38°—42° C., and add 5 c.c. of 20 per cent. lead acetate solution [D].

5. Cool to 15° C. and allow to stand 30 minutes.

6. Decant off alcohol through filter, bring lead soaps on to filter with ether; place filter and lead salts in a Soxhlet extractor, and exhaust with ether till washings are no longer darkened on shaking with SH_2 water.

7. Wash soaps into a separator with an ether wash bottle, and decompose by adding 20—25 c.c. HCl [B] and shaking thoroughly till lead chloride ceases to separate.

8. Run off lower layer and wash ether till free from lead chloride; add a few granules of fused calcium chloride, and stand 30 minutes; run off into flask, rinsing out separator with ether, and retaining any drops of water in the separator.

9. Distil off ether, and dry fatty acids at 100° C.

10. Dissolve fatty acids in 50 c.c. of 90 per cent. alcohol [E]. Cool to 15° C., collect crystals on a small filter, or Gooch crucible, and wash with three successive lots of 10 c.c. of 90 per cent. alcohol (returning the washings each time once or twice to ensure saturation), each time at the same temperature.

11. Wash with 70 per cent. alcohol (rejecting washings) till the filtrate remains clear on adding water. Dissolve crystals in ether, running solution into tared flask, evaporate, dry at 100° C., and weigh. The correction for the number of c.c. of 90 per cent. alcohol used (normally 50+10+10+10=80 c.c.) is added to the weight obtained (Table I above), and the m.p. taken of the fatty acids (page 39).

This should be 71°—72·5° C.   If below 71°, the treatment (from par. 10) must be repeated.

12. Pure arachis oil gives 4·8 per cent. of mixed arachidic and ligno-ceric acids by this method.

∴ Arachis oil present = per cent. weight fatty acids × 21.

Example

Weight of fatty acids obtained   ...   .   0·2975 grm.
C.c. of 90 per cent. alcohol used   ...   .   80 c.c.
Temperature of alcohol   ...   .   17·5° C.

∴ Correction to be added for alcohol (Table I) = ·064 × $\frac{80}{100}$ = ·0512 grm.
Corrected weight   ...   .   0·3487 grm.

∴ **Arachis oil per cent.** = 0·3487 × 21 = **73·2 per cent.**

Olive Oil

Mazzaron's Test

This test, recently proposed[1], is a modification of the Maumené test (page 100) in which the sulphur dioxide gas liberated by the reaction of the oil with sulphuric acid is led into N/10 Iodine solution, and the amount reduced determined.

20 c.c. of oil is treated with 5 c.c. of pure sulphuric acid of s.G. 1·8417. The SO_2 liberated is aspirated through N/10 Iodine solution and the reduction ascertained by means of thiosulphate.   Mazzaron states that olive oil gives a reduction at 20° C. of 2·2 to 2·6 c.c. of N/10 Iodine; other vegetable oils have much higher values, e.g.

Arachis 7, Rape 15, Sesamé 50, Maize 65, Cotton 138, Soya bean 223.

Test for Carbon Disulphide

(" *Sulphur Olive Oils*")

As for the *Halphen* test (see p. 133), using a drop of pyridine, kapok oil, and a small quantity of flowers of sulphur.   On heating, a *crimson coloration* develops in the presence of as little as 0·02 per cent. of carbon disulphide.

CLASS VI

GROUP I

Cacao Butter

BJÖRKLUND'S[2] TEST

Procedure

1. Place about 3 grms of the fat in a test-tube and add twice the weight of ether (or × 2·84 c.c.) at 18° C.   Shake, and warm if necessary till complete solution takes place.

[1] *Staz. Sperim. Agrar. Ital.* 1915, **48**, 583—594.
[2] *Zeit. f. analyt. Chem.* **3**, 233, modified by Lewkowitsch, *Journ. Soc. Chem. Ind.* 1899, 557.

2. Cool in ice and water (o° C.) and note the number of minutes before turbidity or deposition of crystals occurs, and the temperature at which the solution again becomes clear on removal from the ice.

3. Note also the character of the deposited crystals. Cacao butter separates in tufts of crystals on bottom and sides of tube. If 5 per cent. or over of tallow is present, a flocculent precipitate separates from the solution, and the latter appears *turbid* instead of clear, as with genuine cacao butter.

4. Björklund gives the following table :

	Turbidity at o° C. after minutes	Clear solution at ° C.
Pure cacao butter	10—15	19—20
Cacao butter + 5 °/o beef tallow	8	22
Cacao butter + 10 °/o beef tallow	7	25

Remarks

Bolton and *Revis* have found that the test does not detect the addition of some recent cacao butter adulterants, as e.g. Borneo Tallow.

GROUP III. COCONUT OIL GROUP

SHREWSBURY-KNAPP METHOD

Definition

The Shrewsbury-Knapp method, by determining the solubility of the insoluble acids from a fat in alcohol of definite strength, gives a measure of the fatty acids intermediate in molecular weight (chiefly lauric and myristic) between the normal insoluble acids (stearic, etc.) and the soluble acids (butyric, caproic, etc.).

Outline of Method

5 grms of the fat are saponified, the resulting soap decomposed with mineral acid, and the liberated fatty acids dried, treated with alcohol of definite strength, filtered, and the soluble fraction estimated by titration of the alcoholic filtrate with N/10 alkali.

In the modified method of *Elsdon* (see below) the residual fatty acids from the Reichert-Polenske process are employed.

Remarks

Before the authors of this process published[1] their method, similar suggestions had been made by *Vandam*[2] and *Fendler*[3]. It has particular reference to coconut and palm kernel oils, and the less known oils of this group, since these oils consist preponderantly of the glycerides of lauric and myristic acids.

[1] *Analyst*, 1910, **35**, 385.
[2] *Ibid.* 1901, **26**, 320.
[3] *Arb. a. d. Pharm. Inst. d. Univ. Berlin*, 1908.

The process was published in 1910, and was first criticised[1] as yielding no information not already obtainable by the Reichert-Polenske method. Further, other observers[2] were unable to obtain the same results as the authors. Recently *Elsdon*[3] and *Bagshawe* examined the causes for the divergence of the figures obtained, and introduced minor modifications which they stated ensured concordant results. Later, Elsdon[4] showed that the residual fatty acids from the Reichert-Polenske could be used, and the method treated as an extension of this process.

The chief condition for consistent results lies in the accurate adjustment of the strength of the alcohol employed.

In careful hands, the process should form a valuable check on the Polenske figure for this class of oils, since it gives a measure of the most characteristic acids of the group. Further, as shown[5] by Elsdon, when considered in connection with the Polenske value, it yields a clue as to whether palm kernel or coconut is present in a fat, and it thus appears desirable that the test should be performed in conjunction with this determination. The original form of the test is therefore omitted here.

Procedure (Elsdon's modified method)

1. Remove the flask containing the residual fatty acids from the Reichert-Polenske process from the condenser, and cool the contents in water until the acids have set to a solid cake.

2. Break up the cake with a stirring rod, and pour the contents of the flask on to a fine wire gauze sieve. Wash flask and fatty acids with 50 c.c. of cold water.

3. When acids have drained, return them to flask, using a thin iron spatula, and taking care to remove all particles from the gauze.

4. Dry flask and contents in the water oven, blowing air through the flask at intervals.

5. Add to the flask 100 c.c. of alcohol, accurately adjusted to EXACTLY ·9200 S.G. AT 15·5° C. (a standardised thermometer reading to $\frac{1}{10}$° should be employed), from a standard pipette, cork the flask, and warm cautiously until the fatty acids have completely dissolved.

6. Cool under tap below 15·5° C., thoroughly shake, and allow to stand in water at 15·5° for half an hour, or until temperature is exactly 15·5°.

7. Filter; titrate 50 c.c. of filtrate with N/10 Caustic Soda after addition of 1 c.c. of 0·2 per cent. phenol phthalein solution (practice is necessary to judge accurately the end point, which is somewhat difficult to observe).

[1] Cribb and Richards, *Analyst*, 1911, **36**, 327, etc.
[2] Revis and Bolton, *ibid.* 1911, **36**, 334, etc.
[3] *Ibid.* 1917, **42**, 72.
[4] *Ibid.* 1917, **42**, 298.
[5] See under *Palm Kernel Oil*.

Results

Elsdon gives the following figures with butter and margarine mixtures :

Per cent. coconut oil	0	10	20	30	40	50	60	70	80	90	100
With butterfat· ...	12·7	14·5	16·5	20·6	24·0	30·4	—	44·4	58·3	76·5	89·8
With oleo margarine	8·9	10·6	13·7	16·9	20·8	25·4	31·8	41·8	53·4	73·2	89·8

CURVE OF RESULTS.

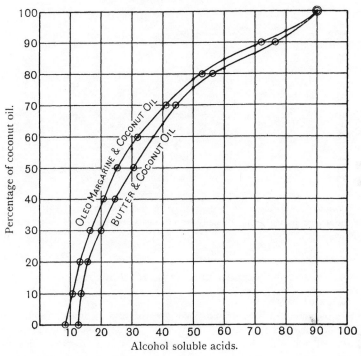

Palm Kernel Oil

The detection of this oil in mixtures where coconut oil is also present is a matter of great difficulty, since apparently the only respect in which these two oils differ chemically is in their content of lower fatty acids, coconut being richer in these acids than palm kernel. This is shown in the higher Polenske value of coconut oil. When coconut oil is pressed, the stearine is found to have practically the same chemical composition as palm kernel oil.

Elsdon[1] has shown that palm kernel oil is indicated in butter and margarine mixtures when the Polenske figure, calculated to coconut oil,

---

[1] *Analyst*, 1917, **42**, 298.

gives a lower percentage than that calculated from the Shrewsbury-Knapp figure, which depends upon the proportion of lauric and myristic acids present, and is practically identical with coconut and palm kernel oils.

Thus, in the figures for commercial samples of margarine below, Elsdon deduces the presence of more or less palm kernel oil occurring together with coconut oil in the case of Nos. 1, 2, 4, 6, 8.

No. of Sample	Reichert	Polenske	S. K. (Elsdon's modification)	Coconut per cent.		Difference
				From P.	From S.K.	
1	6·0	8·2	76·0	52	88	36
2	8·3	6·3	29·0	40	50	10
3	3·2	2·5	11·5	21	18	3
4	12·6	9·3	43·9	54	68	14
5	3·8	2·3	12·5	19	20	1
6	5·2	5·0	27·9	34	46	12
7	5·0	4·7	16·9	33	31	2
8	5·4	6·9	37·5	45	61	16
9	2·4	1·3	8·2	11	10	1
10	5·1	4·7	17·3	32	32	0
11	4·6	3·2	14·1	24	24	0
12	5·4	6·1	22·7	40	41	1

Burnett and *Revis*[1] have devised a method to arrive at the ratio of coconut and palm kernel oils present in fat mixtures. This is given as follows:

Burnett and Revis Method

Outline of Method

In the standard Reichert-Polenske process, after neutralisation of the insoluble volatile fatty acids *with N/10 baryta*, the insoluble barium salts are filtered off, washed with 93 per cent. alcohol, dissolved in a definite amount of hot alcohol of the same strength, and the temperature of turbidity on cooling ascertained.

Remarks

Alcohol of 93 per cent. by volume strength is stated to be the most satisfactory strength to employ, and must be accurately checked. The turbidity point is stated to be quite independent of the amounts of the two fats present, but is determined by their relative percentages.

The test should be made within a few hours of the Polenske titration.

Reagents required

Alcohol of 93 per cent. strength by volume.

[1] *Analyst*, 1913, **38**, 255.

Place 7 c.c. of water at 15·5° C. in a 100 c.c. flask, and make up to the mark with absolute alcohol (Kahlbaum's original purity) at 15·5°. The specific gravity should be $0.8235 \frac{15.5°}{15.5°}$.

Procedure

1. Obtain Polenske figure by standard method (page 116), using N/10 baryta for the titration.

2. Filter the insoluble Ba salts produced on a hardened filter paper, using a filter-pump, and wash with 3 c.c. of 93 per cent. alcohol (above), keeping funnel covered to avoid evaporation.

3. When filter paper is as free from alcohol as possible, place it and the ppt. in a wide-mouthed flask, add *ten times the Polenske value in c.c.* of 93 per cent. alcohol, and boil under reflux condenser till all Ba salts are dissolved.

    When a clear solution cannot be obtained, the flask is left in water at 70° C. until the ppt. has settled, and the liquid then decanted off.

4. Transfer the hot liquid rapidly to a test-tube, enclosed in a larger tube so as to form a jacket, insert thermometer (small bulb) and take temperature of turbidity as in Valenta test (page 79). Warm till just clear and repeat. Take the second temperature as final.

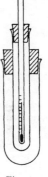

Fig. 52

Results

Burnett and Revis obtained the following results:

Fat	Reichert-Meissl	Polenske	Turbidity temp. °C.
Coconut	7·5	16·5	52·5
Palm Kernel	5·2	9·6	68·5
Palm Kernel oleine	7·2	12·1	59·5
Palm Kernel stearine	—	8·2	72·5
Coconut oleine	—	—	53·0*
Coconut stearine	—	—	63·0*
Coconut oleine, 80 ⎱ Palm Kernel oleine, 20 ⎰	8·24	17·05	54·5
Coconut stearine, 60 ⎱ Palm Kernel stearine, 40 ⎰	4·43	9·93	67·0

* Calculated.

CLASS VII

GROUP I

Lard

Stock-Belfield[1] method for the detection of Beef stearine in lard

Procedure

1. Prepare two sets of standard mixtures.
   A. Pure lard of M.P. 34°—35° C. with
      5, 10, 15 and 20 per cent. of beef stearine, M.P. 56° C.
   B. Pure lard of M.P. 39°—40° C. with
      5, 10, 15 and 20 per cent. of beef stearine, M.P. 50° C.
2. Obtain M.P. (capillary method, page 39) of lard under examination
   Compare this with set A or B, according to which it approximates
   in M.P.
3. Dissolve 3 c.c. of melted fat in 21 c.c. ether in a graduated cylinder
   of 25 c.c. capacity.
4. Dissolve similarly 3 c.c. of each of the standards in the set of
   mixtures chosen.
5. Cool all at 13°C. for 24 hours, and examine the volumes of crystals
   formed.
      The amount of beef stearine is approximately determined by
   comparison of the sample with the standard set.
6. Examine crystals under microscope, and confirm by the characteristic appearance. (See below.)

Microscopic Examination

Crystallise the sample from ether, and examine under a cover-slip
first with ½-inch then with ⅙-inch objective.

Lard crystallises in oblong prisms with CHISEL-SHAPED ENDS.

Beef stearine crystals are tufts of fine needles, but never with chisel-
shaped extremities. (See photos.)

(The original crop of crystals should be examined: after recrystallisation the
characteristic appearance is apt to disappear.)

Method of Bömer[2]

This method is based upon the difference between the melting point of the
glycerides crystallised from lard and that of the fatty acids derived therefrom.

1. Dissolve 50 grms of melted fat in 50 c.c. ether, and allow to stand 1 hour
   at 15°C.
2. Collect crystals on a filter, press, redissolve in 50 c.c. ether, and stand again
   1 hour at 15°C.

[1] *Analyst*, 1894, **19**, 2.
[2] *Zeits. f. Unters. d. Nahrgs- u. Genussm.* 1913, **26**, 559.

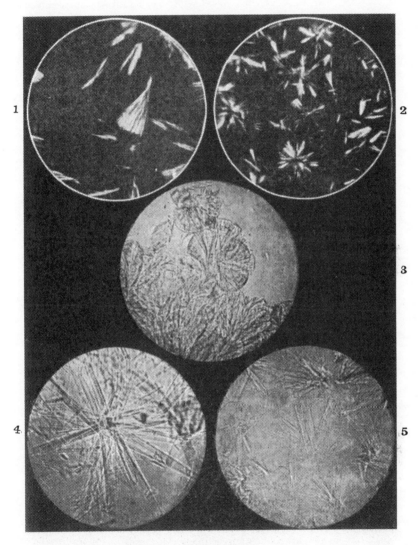

Fig. 53. Beef and Lard Crystals

1. Beef fat × 150   2. Beef fat × 70
(polarised light)   (polarised light)

3. Beef fat × 150

4. Lard × 250   5. Lard × 150

3. Ascertain M.P. of crystals. If below 61° they must be again crystallised from ether.
4. Saponify a portion with alcoholic KOH (page 106), acidifying with mineral acid, and collecting and drying the fatty acids. Ascertain M.P.
5. Lard gives a difference of 5·2°C.
   Beef fat and mutton tallow 0·1 to 2·6°C.
   The presence of beef fat is to be suspected, if, with glycerides giving M.P. 60°—61°, the difference is less than 5·0°C., or with crystals of 65°—68·5° less than 3·0°C.

GROUP II

Butterfat

KIRSCHNER NUMBER

Definition

The Kirschner[1] number is a measure of the volatile fatty acids from an oil or fat, the silver salts of which are soluble in water (e.g. butyric acid in butter).

Outline of Method

The distillate of the water soluble acids in the Reichert-Polenske process, after titration with baryta (N/10), is treated with silver sulphate, filtered, and the filtrate redistilled after addition of dilute sulphuric acid in the standard Reichert-Polenske apparatus, the distillate being measured and titrated as before.

Remarks

This method is an extension of the Reichert-Meissl process, and it indicates what proportion of the soluble volatile fatty acids is due to butyric acid—and therefore to butterfat, since the soluble volatile acids of coconut and palm kernel contain practically no butyric.

In the case of adulteration of butter with coconut oil or palm kernel oil alone, the Polenske value gives all the information required, but when other fats are present, as in the case of mixtures of margarine, containing coconut and (or) palm kernel, with butter the test enables a correct estimate to be made of the total percentage of butterfat present, which could not in many cases be deduced from the Polenske alone. Also, it definitely establishes the presence or absence of small quantities of butterfat in margarines containing these two oils[2].

Procedure[3]

1. Take the filtered distillate after titration with N/10 baryta in the Reichert-Polenske method (care being taken not to exceed the

---

[1] *Ibid.* 1905, **9**, 65.   [2] See page 249.
[3] The procedure and calculation follow those proposed by Bolton and Revis.

neutral point) and place therein 0·5 grm of finely powdered silver sulphate. Insert cork into flask and allow to stand with occasional shaking for 1 hour.

2. Filter; take 100 c.c. of filtrate, place in the standard Reichert-Polenske flask with 10 c.c. of sulphuric acid, dissolved in 35 c.c. of water, adding a stout piece of aluminium wire to prevent bumping.

3. Distil off 110 c.c. as before, in 20 minutes.

4. Titrate 100 c.c. of the distillate with N/10 sodium hydrate, using phenol phthalein as before, and subtract the figure for the blank determination obtained in the Reichert.

5. Calculate Kirschner value as follows:

$$K = x \times \frac{121\,(100 + y)}{10,000},$$

where x = c.c. of K titration, less blank,

y = no. of c.c. of baryta solution used for 100 c.c. of Reichert-Meissl distillate.

Example

Original R.-M. titration,

c.c. N/10 baryta 30·5,

blank, 0·5; corrected, 30·0.

K titration = 15·0; corrected = 14·5.

$$K = \frac{121\,(100 + 30)}{10,000} \times 14·5 = \mathbf{22·6\ Kirschner\ number.}$$

Results

The Kirschner numbers for pure butterfat vary from **19—26**, and, generally speaking, are proportional to the Reichert-Meissl and Polenske values[1]. Coconut gives an average figure of **1·9** and palm kernel oil of **1·0**. Other oils and fats, **0·1—0·2**.

Baryta numbers (Avé Lallement)

Remarks

As shown by *Bolton* and *Revis*[2], the baryta numbers of Avé Lal'ement give useful indications in cases of adulteration with lard and coconut, when both these are present, as both give positive figures, and are not, as in other tests, mutually destructive.

Butterfat always gives a negative value for the formula $b - (200 + c)$, where b = insoluble baryta number, and c = soluble baryta number. In the method the insoluble baryta number is estimated, and the soluble number is obtained by difference from the total baryta number calculated from the saponification value.

---

[1] See Bolton, Richmond and Revis, *Analyst*, 1912, **37**, 183; Cranfield, *Ibid.* 1915, **40**, 439.

[2] *Fatty Foods*, p. 121.

Procedure

1. Saponify 5 grms of fat with 50 c.c. N/2 alcoholic NaOH.
2. Titrate with N/2 HCl, using phenol phthalein. (Calculate S.V., p. 106.)
3. Remove alcohol by boiling and blowing air into flask, and dissolve soap in hot recently boiled distilled water.
4. Transfer to 250 c.c. flask, raise to 40° C., and make up to mark with water at 40° C.
5. Pipette off 100 c.c., stand in boiling water 5 minutes, add 50 c.c. N/5 barium chloride (25 grms $BaCl_2 . 2H_2O$ to 1000 c.c.), and allow mixture to remain in water-bath till Ba salts coalesce.
6. Cool: filter into 250 c.c. flask, washing the insoluble soaps well, and making up to mark.
7. Pipette 200 c.c. into a beaker, acidifying with 1 c.c. conc. HCl: heat to boiling and add 10 c.c. approx. N. H_2SO_4, and stand overnight.
8. Filter ppt. on to a Gooch crucible, wash free from chlorides, and finally with two quantities of 10 c.c. warm alcohol, and dry to constant weight.
9. The weight of $BaSO_4$ found + 25 p.c. is calculated to BaO ($BaSO_4 \times 0.657$); this is subtracted from the BaO value of the $BaCl_2$ solution used in (5) (which must be standardised in an exactly similar way). This gives BaO value of acids forming insoluble salts from 2 grms of fat: and this, calculated to 1 grm fat = insol. BaO value (b).
10. The sapon. value found in (2) is, calculated to BaO (KOH × 1·367), for 1 grm fat = total BaO value (a),

$$a - b = \text{soluble BaO value } (c).$$

11. From these figures[1] calculate

$$b - (200 + c).$$

CLASS IX

Beeswax

Weinwurm's[2] Test (for Ceresin and Paraffin)

Procedure

1. Saponify 5 grms of wax with 25 c.c. of N/2 alcoholic potash. (See page 106.)
2. Evaporate off alcohol completely, and add 20 c.c. pure glycerin, heating on water-bath till complete solution is effected.
3. Run glycerin solution into 100 c.c. of boiling distilled water.
4. Pure beeswax gives *nearly transparent* solution, and print (newspaper, say) can be readily read through 50 mm. of the liquid. 5 per cent. of paraffin or ceresin waxes render print unreadable, and slightly more causes a precipitate.

Remarks

This test depends on the solubility of the unsaponifiable matter of

[1] For results, see under *special tests* in general descriptions of oils and fats in Vol. I.

[2] *Chem. Zeit.* 1897, 519.

beeswax in weak glycerin. Adulteration with carnaüba wax and insect wax also produces turbidity. Although, according to *Berg*[1], some Asiatic beeswaxes give turbidities, it is, on the whole, a good preliminary test.

Hardened fats

ESTIMATION OF MINUTE QUANTITIES OF NICKEL IN FATS

The following method, due to *Atack*[2], is very sensitive and reliable.

Procedure

1. As large a quantity as possible of the fat is placed in a flask and heated on a water-bath with an equal bulk of pure concentrated hydrochloric acid for 1—2 hours, shaking at frequent intervals.
2. The acid is removed by a separating funnel, filtered, and the filtrate evaporated in a porcelain basin on the water-bath (in a fume chamber).
3. The residue is dissolved in 50 c.c. of hot absolute alcohol, rendered just alkaline with ammonia, and 50 c.c. of a hot saturated solution of *a*-benzildioxime in absolute alcohol added. In the presence of Nickel, a red precipitate is produced.
4. The liquid is heated for a few minutes on the water-bath, and then filtered through a tared filter paper, the precipitate being washed with hot alcohol, dried at 110° C., and weighed.
5. The Nickel present is obtained by multiplying the weight of the precipitate obtained by 0·1093.

Example

$$\text{Weight of fat taken} = 253\text{·}57 \text{ grms.}$$
$$\text{Tare of dry filter paper} + \text{weighing tube} = 15\text{·}4325 \text{ grms.}$$
$$\text{,,} \quad \text{,,} \quad \text{,,} \quad + \text{ppt.} \quad = 15\text{·}5273 \text{ ,,}$$
$$\text{Weight of ppt.} = \quad \text{·}0948 \text{ ,,}$$

$$\text{Ni per cent. in fat} = \frac{\text{·}0948 \times \text{·}1093 \times 100}{253\text{·}57} = \text{·}0041 \textbf{ per cent.}$$

Remarks

The precipitate, on boiling, changes from red to reddish yellow. The *a*-benzildioxime is prepared as follows:

1. Dissolve 10 grms of benzil[3] in 50 c.c. methyl alcohol.

---

[1] *Chem. Zeit.* 1903, 753.
[2] *Chem. Zeit.* 1913, **37**, 773; or *Analyst*, 1913, **38**, 316.
[3] For the preparation of benzil and benzoin, see *Pract. Advanced Organic Chem.* (Cohen).

2. Place in a 200 c.c. round-bottomed flask, and add a saturated solution of 8 grms of hydroxylamine hydrochloride in water.
3. Connect to a reflux condenser, and boil for 6 hours on a water-bath, or small direct flame.
4. Filter the precipitate formed, wash it with hot water, and then with a small quantity of cold ethyl alcohol. Dry. (Recrystallise if necessary, from acetone.)

The reaction is capable of detecting 1 in 5,000,000 of Nickel in solution.

The agent originally proposed was dimethyl glyoxime[1].

---

[1] Fortini, *Chem. Zeit.* 1912, **36**, 1461.

SECTION V

IDENTIFICATION AND DETERMINATION OF
FATTY ACIDS AND ALCOHOLS (INCLUDING GLYCERIN)

SECTION V

IDENTIFICATION AND DETERMINATION OF
FATTY ACIDS AND ALCOHOLS (INCLUDING GLYCERIN)

A. MIXED FATTY ACIDS

§ 1. When oils and fats are split by saponification, and the soap so produced acidified by means of mineral acid, each glyceride, simple or mixed, yields its three molecules of fatty acids. If any of the acids so liberated are soluble in water, they are more or less lost in the wash liquors unless special precautions are taken to prevent solution.

The fatty acids therefore normally obtained by the familiar process of saponification of oils, are the so-called INSOLUBLE FATTY ACIDS plus the unsaponifiable matter, and the exact mixture of fatty acids yielded by a particular oil or fat is conveniently termed the **mixed fatty acids** of that particular sample.

Technically, the "fatty acids," prepared by hydrolysis in an autoclave, or by the Twitchell reagent, or otherwise, also consist practically entirely of the fatty acids insoluble in water, the water soluble acids being partly volatilised in the steam used during the deglycerination process, and partly lost by solution in the weak glycerin liquors ("sweet waters").

§ 2. As shown previously[1] fatty acids gradually diminish in solubility as the molecular weight increases. Thus, butyric acid is soluble in water in all proportions, caproic 0·9 in 100; caprylic ·08 in 100; capric 0·1 in 100 boiling water; lauric, traces soluble only in boiling water; myristic, totally insoluble. It is obvious therefore that the liberated fatty acids from fats which contain acids lower in molecular weight than myristic acid will contain indefinite proportions of these lower fatty acids according to the volume of the aqueous mineral acid liquor used, and the length of time of boiling, if any.

The term "mixed fatty acids" refers therefore to the insoluble fatty acids from a particular oil or fat, and for analytical purposes the fatty acids should be well washed with warm water to ensure the removal of soluble acids. It is evident that any reactions or properties dependent on the soluble fatty acids or their glycerides in an oil will be correspondingly modified in the mixed fatty acids derived therefrom.

[1] Vol. I, page 37.

§ 3. The physical and chemical characteristics of the mixed fatty acids resemble closely in most cases those of the oils themselves, except in the case of those properties, such as acidity, melting point, viscosity, etc., which are conditioned by the glycerides *qua* glycerides.

§ 4. Since in the examination of soaps, and of commercial samples of fatty acids, the oils themselves are not available for examination, it will be well to consider the standard determinations from the standpoint of the mixed fatty acids from oils and fats. In the following pages any modifications necessary for the performance of the standard tests are noted, and tables are given showing the figures recorded for the fatty acids by various observers.

Care should be taken in liberating the fatty acids (see page 167) to protect them from oxidation, especially as fatty acids are more sensitive in this respect than the oils from which they are derived. Fatty acids from marine, drying and semi-drying oils should be dried in a current of indifferent gas, such as hydrogen, carbon dioxide, etc.

PHYSICAL.

Specific Gravity (1)

Procedure

Exactly as in the case of oils (page 26) except that, since the fatty acids have usually a higher solidifying point, it is advisable to employ a higher temperature for the determination. The temperature of the boiling-water bath (approx. 99° C.) is the most convenient, since the apparatus can be filled at any temperature and then heated in boiling water, when the excess of fatty acid exudes automatically, leaving the vessel always full.

The S.G. of the mixed fatty acids, taken at 99° (compared with water at 15·5°), differs from that of the oils at the same temperature by being about 0·017 to 0·027 lower, or proportionally less by 2—3 per cent.

The following table shows recorded average specific gravities of the mixed fatty acids of oils and fats taken at $\dfrac{99°\ C.}{15\cdot5°\ C.}$.

Melting Point (2)

The melting point of the mixed fatty acids is much more definite than that of the oils and fats, and is usually higher. The capillary tube method is suitable (page 38). It does not give such good indications as the solidifying point (*q.v.*). Since the M.P. depends upon the proportion of solid acids contained in the oil, and these are often inadvertently or deliberately removed, this figure is very variable. The tables give the average figures only for the fatty acids of each oil.

Solidifying Point (3)

See page 42 and table facing page 70, Vol. I.

TABLE I

Mixed Fatty Acids—Specific Gravities at $\frac{99\cdot0°}{15\cdot5°}$ C.

OILS

VEGETABLE OILS

Class I Marine Oils	Class II Drying Oils	Class III Semi-Drying Oils	Class IV Non-Drying Oils	Class V Animal Oils
Group I Fish Oils — Menhaden, Japan fish, Herring, Salmon	Perilla	Gp. I — Maize 8529, Kapok	Gp. I — Ravison, Rape ·8438, Mustard	Neat's foot ·842*
Group II Liver Oils — Cod liver, Shark liver	Linseed ·8612, Tung, Candlenut, Hempseed, Walnut	Cotton 8467, Sesamé	Gp. II — Apricot kernel, Almond, Arachis ·846, Rice, Olive ·843	Lard oil
Group III Blubber Oils — Seal, Whale ·860*, Dolphin, Porpoise	Soya, Poppyseed ·857*, Niger ·857*, Sunflower	Croton, Curcas	Gp. III — Castor ·896	

FATS

Class VI Vegetable Fats			Class VII Animal Fats	
Group I	**Group II**	**Group III**	**Group I**	**Group II**
Shea	Chinese veg. tallow ·860	Palm kernel	Lard	Butterfat
Mowrah	Palm ·8369	Coconut ·835	Bonefat ·838	
Illipé ·857	Japan wax	Cohune	Tallow	
Cacao	Myrtle wax		Beef ·837*	
Borneo			Mutton	

Note.—Blanks denote no existing records.

* Fryer.

TABLE II
Mixed Fatty Acids—Melting Points °C.

OILS

Class I — MARINE OILS

Group I Fish Oils	Group II Liver Oils	Group III Blubber Oils
Menhaden	Cod liver 20—23	Seal 22—23
Japan fish	Shark liver	Whale 24—27
Herring		Dolphin
Salmon		Porpoise

VEGETABLE OILS

Class II — DRYING OILS

Perilla	—5
Linseed	17—24
Tung	38—40
Candlenut	20—21
Hempseed	18—19
Walnut	16—20
Soya	26—29
Poppyseed	20—21
Niger	28
Sunflower	22—24

Class III — SEMI-DRYING OILS

Maize	18—21
Kapok	32—34
Cotton	34—38
Sesamé	23—26
Croton	
Curcas	24—26

Class IV — NON-DRYING OILS

Gp. I	Ravison	
	Rape	18—19
	Mustard	16—17
Gp. II	Apricot kernel	12—14
	Almond	13—14
	Arachis	31—32
	Rice	32—36
	Olive	24—28
Gp. III	Castor	13

Class V — ANIMAL OILS

Neat's foot	29—31
Lard oil	

FATS

Class VI — VEGETABLE FATS

Group I		Group II		Group III	
Shea	49—52	Chinese veg. tallow		Palm kernel	26—29
Mowrah		Palm	48—50	Coconut	23—27
Illipé		Japan wax	56—60	Cohune	25—29
Cacao	50	Myrtle wax	47—48		
Borneo					

Class VII — ANIMAL FATS

Group I		Group II	
Lard	43—47	Butterfat	38—40
Bonefat	40		
Tallow			
Beef	43—47		
Mutton	48—52		

Refractive Index (4)

The refractive indices of the fatty acids are lower than those of the oils from which they are derived[1]. One advantage in using the fatty acids in preference to the oils for the determination is that no correction is necessary for free acidity, also the latter are sometimes not available. In such cases a higher temperature than 40° C. may be necessary if the S.P. is higher than this ; the results should then be obtained at the M.P. of the acids and calculated to 40° C.

Viscosity (5)

The viscosity of the fatty acids is much lower in a liquid condition than that of the oils at the same temperature. As the figures give no useful indications they have not been recorded.

Solubility[2] (6)

The fatty acids differ greatly in their solubility from the oils from which they are derived, especially in alcohol, in which almost all mixed fatty acids are soluble in the cold, as well as in other organic solvents (castor oil acids are exceptions as regards petroleum ether). Fatty acids are separated by their differing solubilities in alcohol of varying strengths. (See estimation of arachidic, stearic, etc., and Shrewsbury-Knapp Test.)

Polarimeter (7)

Optical activity may be due either to the presence of asymmetric carbon atoms in the glycerides, or in the accompanying sterols, or to small quantities of extractive matters in the oil. In the latter case, the fatty acids will show no optical activity, while in the former case they will give a similar degree of rotation. Thus castor, chaulmoogra, hyndocarpus and other oils give fatty acids showing optical activity.

CHEMICAL.

Iodine Values (8)

Since, so far as iodine absorption is concerned, the glyceryl radicle is a diluent, the fatty acids give higher iodine values than the corresponding oils, but the difference is not constant owing to the slightly differing molecular weights of the fatty acids in combination. The average variation appears to be from 2 to 3 per cent.

Elaïdin (9)

This test gives similar results with fatty acids as for the oils from which they are derived, the oleic acid being converted into elaïdic acid.

[1] The authors find that the correction for each one per cent. free fatty acidity varies from ·000095 to ·000105 for most oils, being higher for coconut and palm kernel and lower for rape and castor. For Butyro-refractometer readings see Table III.
[2] See pages 254, 255 and Appendix.

TABLE III

Mixed Fatty Acids—Butyro-Refractometer at 40·0° C.*

OILS

Class I — MARINE OILS

Group I Fish Oils	Menhaden	53—57
	Japan fish	42—48
	Herring	
	Salmon	54
Group II Liver Oils	Cod liver	50—55
	Shark liver	
Group III Blubber Oils	Seal	48—50
	Whale	41—45
	Dolphin	
	Porpoise	

Class II — DRYING OILS

Oil	Value
Perilla	
Linseed	57—59
Tung	
Candlenut	50
Hempseed	58
Walnut	49—52
Soya	47—48·5
Poppyseed	48—49
Niger	48—49
Sunflower	45—48

Class III — SEMI-DRYING OILS

Oil	Value
Maize	44—45
Kapok	
Cotton	43—44·5
Sesamé	44—45·5
Croton	
Curcas	

Class IV — NON-DRYING OILS

Gp. I	Ravison	
	Rape	46—47
	Mustard	
Gp. II	Apricot kernel	41·5—42·5
	Almond	41—43
	Arachis	
	Rice	41—43
	Olive	
Gp. III	Castor	55—57

Class V — ANIMAL OILS

Oil	Value
Neat's foot	39—40
Lard oil	

FATS

Class VI — VEGETABLE FATS

Group I	Shea	
	Mowrah	34
	Illipé	
	Cacao	33—36
	Borneo	
Group II	Chinese veg. tallow	29—31
	Palm	28—31
	Japan wax	32—34
	Myrtle wax	
Group III	Palm kernel	19·5—21·5
	Coconut	18—20
	Cohune	

Class VII — ANIMAL FATS

Group I	Lard	35—38
	Bonefat	
	Tallow	
	Beef	33—35
	Mutton	33—35
Group II	Butterfat	30—32

* Fryer and Weston.

TABLE IV

Mixed Fatty Acids—Iodine Values (Average)

OILS

Class I — MARINE OILS

Group I Fish Oils		Group II Liver Oils		Group III Blubber Oils	
Menhaden	163	Cod liver	144—174	Seal	143
Japan fish	120—180	Shark liver		Whale	120
Herring	130			Dolphin	
Salmon	169			Porpoise	

Class II — DRYING OILS

Perilla	211
Linseed	179—210
Tung	168
Candlenut	167
Hempseed	149
Walnut	150
Soya	132
Poppyseed	139
Niger	
Sunflower	134

Class III — SEMI-DRYING OILS

Maize	122
Kapok	113
Cotton	105—112
Sesamé	106—111
Croton	111
Curcas	105

Class IV — NON-DRYING OILS

Gp. I	Ravison	126
	Rape	102
	Mustard	108
Gp. II	Apricot kernel	103
	Almond	100
	Arachis	94
	Rice	97
	Olive	86
Gp. III	Castor	89

Class V — ANIMAL OILS

Neat's foot	72
Lard oil	87

FATS

Class VI — VEGETABLE FATS

Group I		Group II		Group III	
Shea	56	Chinese veg. tallow	24	Palm kernel	14
Mowrah	64	Palm	55	Coconut	9.7
Illipé	60	Japan wax	8	Cohune	
Cacao	38	Myrtle wax			
Borneo	32				

Class VII — ANIMAL FATS

Group I		Group II	
Lard	64	Butterfat	35
Bonefat	54		
Tallow			
Beef			
Mutton			

Saponification Value (10)

The corresponding value to this is the "neutralisation value" of the fatty acids. This is performed and calculated in an exactly similar manner to the acid value (page 123), *q.v.*

From this figure is calculated the "mean molecular weight of the fatty acids," thus:

$$M = \frac{56\cdot1}{n},$$

where M = mean molecular wt. of fatty acids,

n = no. of grms of KOH neutralising 1 grm of the fatty acids.

Insoluble Bromide Value of Fatty Acids (11)
See page 109.

Reichert-Meissl and Polenske (12)

The Reichert-Meissl value is of course inapplicable to the fatty acids, but a corresponding figure is the percentage of "**Insoluble acids and unsaponifiable**" (**Hehner number**). This figure is useful technically for calculating the yield of fatty acids to be derived from splitting various oils and fats. The following are directions for performing the test.

Hehner Number

Procedure

1. Saponify 3—4 grms of oil or fat as directed under *sapon. value* (p. 106), rinse the alcoholic soap solution with boiling water into a porcelain dish and evaporate on a water-bath to a thick paste.
2. Dissolve the soap in 100—150 c.c. water and add sulphuric acid 10 °/₀ till liquid contains distinct excess.
3. Heat until the liberated fatty acids float on top as a clear oily layer; weigh a stout, dried, filter paper in a beaker and place in a funnel, filling filter nearly full with hot water.
4. Transfer contents of basin to filter, keeping latter full till all liquid is added. Wash fatty acids with boiling water till washings are no longer acid to litmus (2—3 litres should be employed for butterfat or coconut group of oils).
5. Immerse filter and funnel in a vessel of cold water so arranged that water outside and acids inside are at same level.
6. Drain off water, transfer filter to beaker in which it was weighed and dry at 100° C. Dry for 2 hours and weigh, and for further 1½ hours and weigh again.

   [In case oil belongs to marine, drying or semi-drying group, the fatty acids must be protected from air as much as possible. The alcohol is

TABLE V
Mixed Fatty Acids—Mean Molecular Weights (Average)

OILS

Class I Marine Oils	Group I Fish Oils	Menhaden	289
		Japan fish	
		Herring	292
		Salmon	
	Group II Liver Oils	Cod liver	288
		Shark liver	
	Group III Blubber Oils	Seal	
		Whale	292
		Dolphin	290
		Porpoise	
Class II Drying Oils		Perilla	284
		Linseed	283
		Tung	297
		Candlenut	
		Hempseed	
		Walnut	
		Soya	281
		Poppyseed	
		Niger	280
		Sunflower	
Class III Semi-Drying Oils		Maize	283
		Kapok	293
		Cotton	275
		Sesamé	286
		Croton	285
		Curcas	
Class IV Non-Drying Oils	Gp. I	Ravison	
		Rape	320
		Mustard	317
	Gp. II	Apricot kernel	289
		Almond	
		Arachis	282
		Rice	289
		Olive	286
	Gp. III	Castor	306
Class V Animal Oils		Neat's foot	
		Lard oil	

FATS

Class	Group	Fat	Mean Mol. Wt.
Class VI Vegetable Fats	Group I	Shea	
		Mowrah	
		Illipé	292
		Cacao	295
		Borneo	
	Group II	Chinese veg. tallow	274
		Palm	273
		Japan wax	263
		Myrtle wax	243
	Group III	Palm kernel	223
		Coconut	211
		Cohune	
Class VII Animal Fats	Group I	Lard	278
		Bonefat	280
		Tallow	
		Beef	281
		Mutton	281
	Group II	Butterfat	260

TABLE VI

Mixed Fatty Acids—Insoluble Acids + Unsaponifiable %. [Hehner Value] Average

OILS

Class I MARINE OILS		Class II DRYING OILS		Class III SEMI-DRYING OILS		Class IV NON-DRYING OILS			Class V ANIMAL OILS	
Group I Fish Oils		Perilla	95·9	Maize	95·7	**Gp. I** Ravison			Neat's foot	
Menhaden	95·5	Linseed	95·5	Kapok	94·9	Rape	95·1			
Japan fish	95·0	Tung	95·0	Cotton	95·9	Mustard	95·3			
Herring	95·6	Candlenut	95·5	Sesamé	95·8	**Gp. II** Apricot kernel	95·4			
Salmon		Hempseed		Croton	89·1	Almond			Lard oil	
Group II Liver Oils		Walnut	95·4	Curcas	95·5	Arachis	95·3			
Cod liver	95·7	Soya	95·5			Rice	95·6			
Shark liver		Poppyseed	95·3			Olive	95·1			
Group III Blubber Oils		Niger	94·1			**Gp. III** Castor				
Seal	95·7	Sunflower	95·0							
Whale	95·3									
Dolphin										
Porpoise										

FATS

Class VI VEGETABLE FATS

Group I		Group II		Group III	
Shea	96·5	Chinese veg. tallow	95·6	Palm kernel	91·1
Mowrah	94·7	Palm	89·8	Coconut	89
Illipé		Japan wax	89·7	Cohune	
Cacao	94·6	Myrtle wax			
Borneo	93·3				

Class VII ANIMAL FATS

Group I		Group II	
Lard	95·8	Butterfat	87·5
Bonefat			
Tallow			
Beef	95·6		
Mutton	95·5		

evaporated off in the original flask, and the soap decomposed in the flask. The fatty acids are washed on the filter, keeping latter covered as much as possible with watch glass, and acids are then rinsed with ether into tared flask and ether evaporated off in current of H or CO_2.]

$$\frac{\text{Final weight of fatty acids}}{\text{weight of oil taken}} \times 100 = \text{"HEHNER NUMBER."}$$

The POLENSKE values, obtained by taking 5 grms of the mixed fatty acids in place of the oils, differ little from the figures for the oils themselves. Thus the insoluble fatty acids from coconut oil give a figure of 17·5, and from palm kernel 11·0. The method is of great value in the determination of these oils in soaps, the mixed fatty acids being first obtained therefrom by acidifying the soap solution with mineral acid[1].

Acetyl Value (13)

The acetyl value of the fatty acids can be determined in the same manner as that of the oils, but care should be taken to thoroughly wash the mixed acids free from soluble acids. The method is applicable for the determination of castor oil in soaps.

Acid Value (14)

See *Neutralisation Value* above (p. 167).

Unsaponifiable (15)

Performed in a similar manner to the method for oils (page 126). The fatty acids are first neutralised with potash or soda, and the unsaponifiable matter then shaken out with ether.

Commercial fatty acids frequently contain excess of unsaponifiable owing to decomposition taking place during splitting in autoclaves, etc., or especially in the case of distilled fatty acids. (For soapmaking purposes fatty acids should not exceed 2 °/₀ unsaponifiable.)

[1] See Fryer, *Jour. Soc. Chem. Ind.* 1918, **37**, 262.

B. SEPARATION AND DETERMINATION OF THE FATTY ACIDS

Unsaturated Fatty Acids

All oils and fats give iodine absorption values, and therefore contain glycerides of unsaturated acids (see chapter III, Vol. I). The degree of unsaturation of the acids present is not necessarily determinable by the iodine value, but, in point of fact, it gives an excellent indication, since acids of the highest degree of unsaturation occur, generally speaking, in oils with the highest iodine values. Thus, no oils are known consisting of, e.g., linolenic acid and a saturated acid only.

The complete separation of the unsaturated acids is not possible, but use is made of the differing solubility of the *lead salts* in *ether* to effect an approximate separation in most cases. By means of their lead salts the fatty acids can be divided as follows :

(1) SOLID SATURATED ACIDS—Lead salts *insoluble* in ether, as palmitic, stearic, etc.

(2) SOLID UNSATURATED ACIDS— Lead salts *partially soluble* in ether, as erucic, iso-oleic, etc.

(3) LIQUID UNSATURATED ACIDS—Lead salts *soluble* in ether, as oleic, linolic, linolenic, clupanodonic.

A fourth class, the *liquid saturated acids*, are soluble in ether in all proportions, and the saturated acids of molecular weight less than myristic, become increasingly soluble as the number of carbon atoms decreases, till the liquid acids are reached.

SEPARATION AND DETERMINATION OF LIQUID UNSATURATED FATTY ACIDS

Lead-salt-ether method

Remarks

This method was originally proposed by Gusserow[1] and by Varrentrapp[2], and has been modified by other observers[3].

Procedure

1. Saponify 5 grms of the oil [as in sapon. value, p. 106].

---

[1] *Liebig's Annal.* 1828, **27**, 153.   [2] *Ibid.* 1840, **35**, 197.
[3] Muter, *Analyst*, 1889, 61; Lane, *Jour. Amer. Chem. Soc.* Feb. 1893; Lewkowitsch, *Chem. Tech. and Anal. Oils*, Vol. I, p. 545; Bolton and Revis, *Fatty Foods*.

2. Add acetic acid, till slightly in excess, and back-titrate to neutrality (using phenol phthalein) with alcoholic potash.

> If the mixed fatty acids are employed instead of the oil, dissolve in alcohol, and titrate with $N/2$ aqueous potash. Dilute with water to 100 c.c.

3. Distil off alcohol on water-bath, and dissolve soap in 100 c.c. water.

4. Place in 500 c.c. wide-necked, conical flask, and add slowly, with constant shaking, 200 c.c. of boiling water to which has been added 30 c.c. of 10 per cent. lead acetate solution. Fill up flask with boiling water, and set aside to cool.

5. When the lead salts are set and the liquid clear, pour off the latter, being careful to return any particles of lead salts which run out. Wash the salts with boiling water, shaking and cooling under the tap, meanwhile rotating (see fig. 50) so as to cause the salts to adhere to the sides of the flask. When cold, run out wash water, and repeat with boiling water.

6. Drain flask by turning upside down, and remove any further moisture with a wad of cotton wool held by the forceps.

7. Add 200 c.c. of ether (methylated), cork flask, and shake thoroughly ; then place under reflux condenser, and heat on water-bath till only a fine light suspension remains. Remove from condenser, cork, and set aside to cool to room temperature, filter into a separating funnel, keeping funnel covered with a large watch glass.

8. Wash out flask with four successive quantities of 30 c.c. of ether on to ppt. and allow all ether to drain through.

9. Add to ethereal solution of soluble lead salts in separating funnel, 100 c.c. of dilute HCl (20 c.c. conc. HCl to 100 c.c. with water), stopper the separator, and shake thoroughly to decompose the lead salts. Stand till a complete separation occurs between the ether and aqueous layers, and then run off precipitated lead chloride and water, and wash ethereal layer with lots of 10 c.c. water till free from mineral acid, separating well each time.

10. Add a few granules of fused calcium chloride to ether, and stand for an hour, then remove stopper, and *pour* out the ether through *neck* of separator on to filter and run filtrate into a 200 c.c. graduated flask. Rinse out separator with two lots of ether, being careful to retain calcium chloride and any drops of water which have separated out.

11. Make up the ethereal solution to 200 c.c. with ether, shake, and stopper flask. Pipette into two flasks, A and B (latter tared), 50 c.c. of the ethereal solution, and drive off ether in a current of CO_2 on a water-bath. Continue to heat and pass in CO_2 till no further loss in weight occurs.

12. Immediately introduce into A the requisite quantity of Wijs' iodine solution, and ascertain iodine value. Weigh flask B.

13. Yield of liquid unsaturated acids

= wt. of fatty acids in flask $B \times 4 \times \frac{100}{5}$ per cent.

The yield of solid acids is obtained by the difference of the above figure and the yield of total insoluble fatty acids from the oil.

The yield of solid acids may be directly obtained by treating the insoluble lead salts left on the filter paper as described on page 137. The iodine value should be determined. The acids should give no ash on incineration, showing that the lead salt has been properly decomposed.

SOLID ACIDS

Separation and Determination

The solid acids are separated and estimated by their differential solubility in alcohol of varying strengths.

BEHENIC⎱ as described in *Tortelli* and *Fortini* test for rape oil
ERUCIC ⎰   (page 135).

ARACHIDIC as described in *Renard's* test (page 144).

STEARIC.

Stearic acid is separated and its percentage estimated in the mixed fatty acids of an oil or fat by treatment of the acids with a saturated solution of stearic acid in alcohol of ·8183 S.G. at 0° C. Under these conditions the liquid acids, and acids of lower molecular weight than stearic are dissolved out. The following method is due to *Hehner* and *Mitchell*[1] and revised by *Emerson*[2].

Estimation of Stearic Acid

Procedure

1. Dissolve 3·5 grms pure stearic acid in 500 c.c. of alcohol S.G. ·8183 (= 94·4 per cent. by volume) in a stoppered bottle, and place in ice and water overnight.

2. Next day, siphon off the supernatant alcohol by means of a thistle funnel, tied over the end with fine muslin, and connected with a vacuum flask and filter pump (fig. 54).

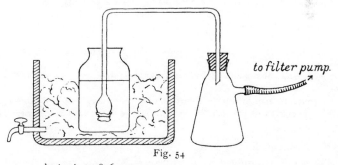

to filter pump.

Fig. 54

[1] *Analyst*, 1896, 321.
[2] *Zeits. f. Unters. d. Nahrgs- u. Genussm.* 1903, 6, 22.

3. Weigh into a squat wide-mouthed flask 1 grm of solid (or 5 grms if liquid) acids, and dissolve in 100 c.c. of above alcohol solution. Cool to 0° in ice overnight, and agitate to promote crystallisation.

4. Filter off alcohol as before, keeping flask immersed in ice water, using thistle funnel and vacuum flask. Wash residue with three lots of 10 c.c. alcoholic stearic solution at 0° C.

5. Dissolve crystals adhering to muslin of thistle funnel with ether and add to flask, evaporate off dry at 100° and weigh.

6. Ascertain M.P.: should be not under 68·5° C.: if too low the treatment should be repeated.

> (In order to test if free from fatty acids of higher mol. weight, titrate with standard alkali (page 124).)

7. Subtract from weight of crystals obtained ·005 grm to allow for stearic acid due to alcoholic solution used. The balance of crystals is calculated to *stearic acid per cent.*

Palmitic acid cannot be satisfactorily estimated in presence of other solid acids and is usually obtained by difference, or by fractional crystallisation.

LIQUID ACIDS

Separation and Determination

The liquid fatty acids are first separated from the solid acids by the lead-salt-ether method (page 171).

They are then brominated, as described under *ether insoluble bromine value* (page 109). The precipitated bromides are determined (if any) and their M.P. ascertained. If they melt to a clear liquid at or about 180° C. there is no *octobromide* present (from clupanodonic, or similar acid).

Octobromides blacken without melting at 200° C.

In the case of mixtures of octo- and hexabromides (both of which are insoluble in ether), the latter may be separated from the former by treatment with pure boiling benzene, in which the *hexabromide* is the more soluble.

The mother liquor from the bromination treatment contains an excess of bromine. This is removed by placing the ethereal solution in a separating funnel and shaking with an aqueous solution of sodium thiosulphate, and the water in the ethereal solution removed by filtration through a dry filter paper.

The ether is distilled off, and the residue boiled with a minimum of petroleum ether to completely dissolve it (B.P. of petroleum ether below 40° C.). On cooling, crystals of bromide melting at 114° C. separate out. This is the *tetra-bromide.*

On concentrating the filtrate, a further crop of crystals may be obtained, and the filtrate consists of the *dibromide* which may be determined by evaporation of the solvent.

Hydroxylated Fatty Acids

These are determined by the *acetyl value* (page 119) of the mixed insoluble acids after washing free from soluble acids.

Oxidised Fatty Acids

Oxidation of the oils and fats takes place on exposure to air and light, the change not being thoroughly understood.   Resinous bodies are produced which are distinguished by their *insolubility* in *petroleum ether*.   Such "oxidised" fatty acids differ in this way from hydroxylated fatty acids and polymerised acids, though the latter are often formed side by side with the oxidised bodies.   Their determination is carried out as follows:

Procedure

1. Saponify 5 grms of the oil with alcoholic potash (page 106).
2. Evaporate off alcohol, dissolve soap in hot water, place in separating funnel, and decompose by adding hydrochloric acid.
3. Cool: shake with petroleum ether (B.P. 40°—80° C.) and stand until ethereal layer completely separates.
4. The oxidised acids adhere to sides of flask in the petroleum ether layer: run off aqueous liquid, and run the ethereal layer through filter paper. Wash the oxidised acids with petroleum ether to remove adhering fatty acids.   Then dissolve in warm ether (methylated), transfer to tared flask, washing out filter: dry and weigh.

   If the amount of oxidised acids is considerable, resaponify with alkali and repeat operations.

C. EXAMINATION OF ALCOHOLS

Alcohols exist in oils, fats and waxes in three forms :

I. In oils and fats as the aromatic alcohols known as the **sterols**, which are characteristic of natural fats.

II. In most natural waxes, **as aliphatic, monohydric alcohols**, combined with fatty acids as esters, and often free in the wax.

III. As the trihydric base **glycerol** in combination with fatty acids in all oils and fats.

I. Sterols

As previously explained (Vol. I, page 62) the sterols are contained in all natural oils and fats, CHOLESTEROL being characteristic of all oils and fats of animal origin, and SITOSTEROL, or other phytosterol, of fatty bodies of vegetable origin.

The amount of sterol present in oils and fats varies usually from 0·1 to 2·0 per cent., and it is contained in the unsaponifiable matter, isolated from the soap solution of the oil by solvents. The sterols are conveniently estimated by the method of *Windaus*, by precipitation with an alcoholic solution of DIGITONIN, which forms insoluble compounds with sterols.

Estimation of the Sterols

Procedure

1. Dissolve the unsaponifiable matter in 50 parts of boiling 95 per cent. alcohol (if not clear, filter, and wash with little alcohol), add hot 1 per cent. digitonin solution in 90 per cent. alcohol till no further precipitation takes place.

2. Settle for a few hours, filter on to a Gooch crucible, wash with alcohol, then with a little ether, dry at 100° C. and weigh ; weight of precipitate × 0·25 gives amount of sterol present.

Remarks

If using the original oil, leach 50 grms of the filtered oil or fat with 75 c.c. of 95 per cent. alcohol, boiling under reflux, cooling, and separating off alcohol ; then treating oil similarly with a further 75 c.c. of fresh alcohol. Treat the alcohol extracts with 1 per cent. digitonin in hot 90 per cent. alcohol as before.

IDENTIFICATION OF THE STEROLS

There are two methods of identifying the sterols, viz. by microscopic examination of the crystals formed from alcoholic solutions, and by ascertaining the melting points of esters, of which the acetates are the most suitable.

Since the presence of phytosterol in an animal fat postulates admixture with a vegetable oil or fat, the recognition of the sterols is of great analytical importance. It is comparatively easy to detect phytosterol in animal fats, but the recognition of cholesterol in vegetable oils and fats is much more difficult; a commonly occurring instance of this is the adulteration of linseed oil with fish oils.

Microscopic Examination

Procedure

1. Dissolve unsaponifiable matter (page 126) in a small quantity of cold ether and run into a small porcelain dish. Evaporate off solvent spontaneously, dry on water-bath, cool, and dissolve in the minimum amount of absolute alcohol.

2. Run the clear decanted solution into a small test-tube, and allow to crystallise by spontaneous evaporation. Obtain a sample from the bottom of tube by means of a fine pipette, place on cover glass, and examine under microscope.

    If crystals are not observed, purify the warm alcoholic solution by shaking with finely powdered animal charcoal; filter warm, evaporate to dryness and redissolve in a little alcohol.

3. Cholesterol gives crystals as in figure. They appear as plates, rhombic in form (triclinic system).

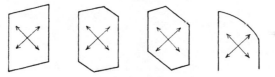

Fig. 55. Cholesterol crystals

Phytosterols give needles in star-shaped groups, and crystals of the form shown are obtained.

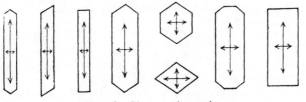

Fig. 56. Phytosterol crystals

Mixtures of the sterols give the phytosterol form, unless cholesterol is in excess, when needles and rhombic plates are found side by side; also telescopic forms occur as shown in figure.

Fig. 57.   Telescopic crystals from mixture of sterols

The microscopic examination is only preliminary, and should be followed by the " Phytosteryl acetate test " (under).

Phytosteryl Acetate Test

Procedure

1. Take the alcoholic solution of the crystals (§ 2, p. 177) and evaporate to dryness in a small porcelain dish, then add 2—3 c.c. of acetic anhydride (per 100 c.c. original oil), cover dish with watch glass and heat to boiling with a small flame.

2. Evaporate off excess of acetic anhydride on the water-bath, add a small quantity of absolute alcohol; warm, and set aside to crystallise.

3. Filter crystals and wash with little 95 per cent. alcohol; transfer to a small basin, dissolve in 5—10 c.c. absolute alcohol and allow to recrystallise. Place a few crystals on a porous plate to dry, and take M. P. (page 38).

4. Recrystallise the remainder of the crystals three times by filtering and redissolving in alcohol, taking a small portion of the crop produced each time. Ascertain M. P. of each crop.

5. Crystals are now obtained from 2, 3, 4, 5 crops. If the M. P. of 5 does not differ from 4, no further crystallisation is necessary. If it varies, further crystallisation is advisable.

6. If the M. P. of crop 5 is below 115° C., phytosterols are absent. If, on the other hand, the M. P. is higher than this, and tends to rise with each recrystallisation, phytosterols are present. The following table gives an idea of the percentage of phytosterol. (*Jäger and Klamroth.*)

Remarks

The digitonin compound may be employed (as obtained previously, p. 176), the dry ppt. being boiled with acetic anhydride.

If the crystals of the first crop are coloured, they should be pressed between filter paper and, if necessary, decolorised by shaking the warm alcoholic solution with finely ground animal charcoal, and filtering hot.

Should paraffin wax be present (as adulterant) the melting points of the crystals will be greatly lowered. The wax must in this case be first removed by the method of *Polenske*.

Treat the unsaponifiable matter from 100 grms of fat with 1 c.c. of petroleum ether (B. P. below 40° C.) for 20 minutes at 15° C., transfer to small funnel closed with a plug of cotton wool, and wash with five successive portions of ½ c.c. petroleum ether. Cholesterol and phytosterol remain on the plug, the paraffin is dissolved. The residue is then acetylated as before. The paraffin wax may be estimated by evaporating off the petroleum ether filtrate and washings and weighing, or in the following manner :

The unsaponifiable matter from 100 grms of fat is treated with 5 c.c. of conc. H_2SO_4 in a glycerin bath at 104—105° for 1 hour ; the sterols are thus destroyed and the wax remains : extract wax with petroleum ether and weigh.

Sterols per cent.		Crystals M.P.°C.	
Cholesteryl Acetate	Sitosteryl Acetate	Jäger	Klamroth
100	—	113·6	113·0
96	4·0	115·0	114·2
86·4	13·6	119·1	118·0
79·5	20·5	121·3	120·6
66·6	33·4	125·0	124·2
50·0	50·0	127·1	127·0
38·3	61·7	128·2	127·8
33·3	66·7	128·3	128·2
26·4	73·6	128·3	128·0
15·5	84·5	128·2	128·0
10·1	89·9	128·1	127·8
0	100·0	127·1	127·0

Pure Cholesteryl acetate melts at 114·3—114·8° (corrected).
Phytosteryl acetates from different oils melt *above 125°*C.

The figures of the table should be compared with the M. P. *of the 2nd crop of crystals, but this should be confirmed by the recrystallisations described.*

II. **Wax Alcohols**

The alcohols contained in the waxes are either free or in combination with fatty acids. They almost all belong to the higher members of the monohydric series (see Vol. I, p. 58). As hydrocarbons are also frequently present in waxes, these are obtained, together with the alcohols, on determining the "unsaponifiable" matter of waxes by the usual method (page 126). An estimate of the alcohols may be obtained by means of their ACETYL VALUE (page 119). The following are the acetyl values of the commoner wax alcohols :

Cetyl	alcohol		197·5
Octodecyl	,,		180·0
Ceryl	,,		128·1
Myricyl	,,		116·7
Sterols			135·5
Mixed alcohols :	sperm oil		161—190
,,	wool fat		160·9
,,	beeswax		99—103
,,	carnaüba wax	123	

The sterols, if present, and mineral waxes may be separated from the aliphatic alcohols by heating with an equal weight of potash lime (1 part KOH, 2 parts CaO) to 250° C. for two hours. Under these conditions the aliphatic alcohols lose hydrogen and are converted into the salts of the corresponding fatty acids, thus :

$$C_{30}H_{61}OH + KOH = C_{30}H_{59}O_2K + 2H_2$$
$$\text{myricyl alcohol} \qquad \text{potassium melissate}$$

The sterols, remaining unchanged, may be extracted by treating powdered mass in a Soxhlet extractor with ether ; any solid hydrocarbons will also be extracted, and may be separated from the sterols as previously described.

III.　GLYCERIN

Glycerol or glycerin is the characteristic constituent of all natural oils and fats, in which it exists as the glyceryl base $C_3H_5 \Big\langle$ in combination with fatty acids.

It follows that the percentage of an oil or fat present in a mixture with non-glyceridic substances can be approximately estimated by the yield of glycerin obtained on saponification. (See however below.)

The percentage of glycerol present in oils and fats varies on account of the possible presence of varying amounts in the oil of

　　1.　Free fatty acids.

　　2.　Mono- and diglycerides.

　　3.　Unsaponifiable matter.

As shown previously (Vol. I, page 85) the glycerol obtainable from a pure triglyceride mixture can be calculated from the saponification value, each grm of KOH being equivalent to 0·54664 grm of glycerol. The percentage of glycerol yielded by pure triglycerides can therefore readily be calculated from the table of the saponification values given on p. 264, Vol. I.

The theoretical yield from neutral oils and fats varies from about 9·5 (rape) to 14·0 (coconut), being normally about 10·5[1].

In the case of free acidity in an oil a proportionate amount of glycerol is lost owing to the decomposition of the glyceride.

[1] See Table XII.

DETERMINATION OF GLYCERIN IN OILS AND FATS

Procedure

1. Saponify an accurately weighed quantity (20–30 grms) of the oil or fat with alcoholic potash (page 106), and drive off alcohol on the water-bath.
2. Add water to dissolve the soap paste, and then sufficient dilute (1–4) sulphuric acid to decompose the soap.
3. Filter off the fatty acids, wash, and neutralise the excess of acid in the filtrate with an excess of barium carbonate, filter again, wash, and evaporate the liquid on the water-bath till no further water is driven off.
4. Exhaust the residue with small volumes of a mixture of ether and alcohol, washing through a dry filter, and running filtrate and washings into a small beaker, evaporate off solvent in hot water, dry the residual crude glycerin in a desiccator over sulphuric acid, and weigh.

   Drying to constant weight is unnecessary.
5. Determine acetin value of crude glycerin (page 188), and from this calculate the yield of glycerin from the oil.

Example

Oil taken—25·256 grms,
wt. glycerin obtained—3·037 grms,
percentage of glycerol, by acetin process—88·75.

$$\therefore \text{ percentage yield of glycerol from fat} = \frac{3\cdot037 \times 88\cdot75 \times 100}{100 \times 25\cdot256}$$

$$= \mathbf{10\cdot64} \text{ per cent.}$$

COMMERCIAL GLYCERINS

§ 1. Remarks

CHEMICALLY PURE GLYCERIN is obtained by a double distillation of crude glycerins, followed by bleaching with animal charcoal.

DYNAMITE GLYCERIN is the term employed for the once distilled crude. It varies from deep yellow to pale straw in colour.

CRUDE GLYCERINS may be derived from soap lyes ("soap crude") or from the various fat-splitting processes ("saponification crude").

The candle-crude glycerin, derived from the acid saponification process, is termed "distillation crude."

Each type of glycerin has to comply with certain standards, according to the purpose for which it is employed. The following table summarises the principal requirements of the four grades of glycerin.

	Soap crude	Saponification crude	Dynamite	Chem. pure
Glycerol per cent.	80·0	88·0	98—99 °/₀	over 99 °/₀
Ash ,,	10·0	0·5	under 0·25	under ·01
Organic residue	3·0	1·0		under ·05
Specific gravity 15° C.	1·3	1·239—1·243	1·261—1·263	1·260—1·265
Arsenic	traces only	traces only	minute traces only	absent

§ 2. **Chemically pure** glycerin may be tested by its specific gravity (see Table IV, Vol. I), its refractive index (see Table XI) or its viscosity (Table VII, p. 265, Vol. I). It should however conform to the following tests for purity :

Ash. An average sample gives an ash of ·005 per cent., and the ash should not exceed ·01 per cent. The B.P. requirement is "no appreciable ash." The ash will contain copper or iron. The B.P. test for *iron* is to mix 10 c.c. glycerin with 40 c.c. water, add a drop of ammonium hydrate and a drop of a solution of tannic acid, when not more than a *faint* and *transient* pink or purple coloration occurs.

Lead and copper must be absent when tested for as prescribed by the B.P. (comparing colours produced by addition of (1) ammonia and potassium cyanide for copper and (2) hydrochloric acid and H_2S for lead. The solutions are compared in Nessler glasses, one containing $\frac{1}{5}$ the weight of glycerin of the other. No difference in colour should be observed between the two solutions).

Organic Impurities

These may be derived from faulty manufacture, or may be due to adulteration. The *organic residue*, resulting from subtracting the ash from the residue obtained on gently evaporating a weighed quantity of the glycerin at 160° C., does not normally exceed ·04 per cent.[1] The B.P. tests are as under :

Sugar. Only slight charred residue and no odour of burnt sugar when heated with naked flame. (Sugar, if present, may be readily estimated by means of the polarimeter.)

Acrolein, formic acid. Should undergo no darkening in colour when mixed with an equal volume of a solution of ammonia and a few drops of $AgNO_3$ soln. and allowed to stand 5 minutes in the dark.

Fatty acids. Warm gently with an equal volume of dilute sulphuric acid, and shake vigorously. No more than a faint odour should be noticeable.

"**Extraneous organic matter.**" Shake with an equal volume of conc. H_2SO_4 and keep cool. No more than a very slight straw coloration should occur.

Arsenic

This should be entirely absent. The B.P. limit is two parts per million. The standard test is due to *Gutzeit* :

(*a*) Into a long test-tube place 2 c.c. of the glycerin.

(*b*) Add some zinc free from arsenic, and a few c.c. of pure dilute sulphuric acid.

(*c*) Over the top of the test-tube place a cap of filter paper, moistened with a saturated solution of mercuric chloride ; then three further layers of dry filter paper, and secure by means of a rubber band.

(*d*) Examine after half an hour; no stain should be produced on the paper moistened with the $HgCl_2$.

For medicinal purposes, a convenient apparatus for the test is that shown in the figure (*Kirby-Gutzeit*). It is employed in the following manner :

---

[1] Due to polyglycerols. Care should be taken not to overheat, or these will be formed in larger proportions.

(*a*) Dissolve 10 grms of glycerin in 10 c.c. of water in the flask **a**, add 10 c.c. of pure arsenic-free H_2SO_4 and cool the mixture under the tap.

(*b*) Half fill the absorption bulbs **b**, **b** with 5 per cent. lead acetate solution, and tie firmly on to the thistle head **c** of the apparatus a cap of filter paper moistened with 3 drops of 5 per cent. $HgCl_2$ solution and dried.

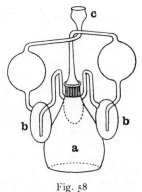

Fig. 58

(*c*) Place 7 grms of arsenic-free zinc in the flask, and fix the upper half of the apparatus in position.

(*d*) If there is no evolution of hydrogen, add water to the flask in small quantities (1 c.c. at a time) until hydrogen is evolved at the rate of about 1 bubble per second.

(*e*) Make further small additions of water so as to maintain this rate of evolution of gas, until all the zinc is dissolved (2—3 hours).

(*f*) Examine the filter paper cap for stain, and compare the depth of colour with that obtained by using known quantities of arsenic in the apparatus.

§ 3. **Distilled Glycerin (Dynamite Glycerin)** has a S.G. of 1·261 to 1·263. The colour varies from deep yellow to pale yellow. The percentage of glycerin can be approximately determined by referring to the S.G. tables.

S.G. not under 1·261 at 15·5° C.

Lime, Magnesia, Alumina should be absent.

Chlorine, traces only permissible (no milkiness with $AgNO_3$).

Arsenic, traces only permitted (Gutzeit test above, p. 182). No yellow ppt. on making the glycerin *just* alkaline with ammonia and adding $AgNO_3$.

Organic impurities. Glycerin must not darken within 10 minutes of adding a few drops of $AgNO_3$ to dilute solution.

Total residue should not exceed 0·25 per cent.

Free acids. Glycerin should not be acid to litmus. Test for volatile acids as above.

Nitration test

The suitability for nitroglycerin manufacture must be definitely ascertained by a direct test. (*Owing to the dangerous nature of this test, it should be restricted to performance with all due precautions by qualified senior chemists.*)

1. Take 50 grms of nitric acid, S.G. 1·5, and pour cautiously with

stirring into 100 grms sulphuric acid, s.g. 1·845, contained in a flask. Cool under tap.

2. Weigh into a thin-walled beaker of about 200 c.c. capacity 112 grms of the above acid mixture, immerse in a mixture of water and ice, and stir with a thermometer (cautiously to prevent fracture of the bulb).

3. When the temperature has fallen to 12° C., weigh out accurately about 15 grms of the glycerin in a beaker with spout, and allow to fall into the acid mixture *drop by drop*, stirring cautiously and not allowing the temperature to rise above 30° C. *After each drop has fallen, the liquid is stirred till the temperature has fallen to 25° before the next drop is added.*

    [If the temperature rises to danger point for any reason, the safest course is to break the bottom of the beaker with the thermometer and discharge the contents into the ice water.]

4. When all the glycerin has been introduced in this way stir the mixture a few minutes longer until the temperature has fallen to 15° C., and transfer to an absolutely *dry* separating funnel (rinsing the funnel with conc. H_2SO_4 is advisable).

5. If the dynamite glycerin is of good quality, the nitroglycerin produced will quickly rise and float as an oily layer on the mixed acids. The quality of the glycerin is judged by the rapidity of the separation into a clean layer. If an intermediate layer of flocculent matter is apparent, or if the separation be slow, or the nitroglycerin contain flocculent material, the glycerin is unsuitable for nitroglycerin manufacture, and must be rejected.

6. While the separation is taking place, re-weigh the beaker used for the glycerin, and calculate the glycerin employed in the test.

7. When complete separation of the nitroglycerin has taken place, run off the acid layer carefully, and slightly agitate the nitroglycerin (without shaking) so as to cause any adherent drops of acid to deposit: draw this off, and then wash with water at 35—40° C., and again twice with 20 per cent. sodium carbonate solution, lastly with water.

8. Transfer nitroglycerin to a burette or cylinder, and read off the volume of nitroglycerin produced.

9. The approximate yield of nitroglycerin is obtained by multiplying the no. of c.c. by 1·6; thus

$$\text{yield} = \frac{\text{No. c.c. nitroglycerin} \times 1\text{·}6 \times 100}{\text{Wt. of glycerin taken}}.$$

The yield should be over 210 per cent. (theoretical yield 246·7 per cent.).

10. *The* nitroglycerin must be destroyed. *Spread out a thick layer of dry sawdust in an open space, away from buildings, and allow the nitroglycerin to drop on to it from a separating funnel in a thin*

continuous trail, taking care that no pools are formed. Apply a lighted match to one end, and the nitroglycerin will burn away rapidly. [Destroy acids in a similar way, but apart from the nitroglycerin.]

CRUDE GLYCERIN. STANDARD METHODS OF ANALYSIS (I.S.M.)

Sampling

The most satisfactory method available for sampling crude glycerin likely to contain suspended matter, or which is liable to deposit salt on settling, is to have the glycerin sampled by an official sampler as soon as possible after it is filled into drums, but in any case before any separation of salts has taken place. In such cases he shall sample with a sectional sampler (page 17), then seal the drums, brand them with a number for identification, and keep a record of the brand number. The presence of any visible salt or other suspended matter is to be noted by the sampler, and a report of same made in his certificate, together with the temperature of the glycerin. Each drum must be sampled. Glycerin which has deposited salt or other matters cannot be accurately sampled from the drums, but an approximate sample can be obtained by means of the sectional sampler, which will allow a complete vertical section of the glycerin to be taken, including any deposit.

Analysis

I. Determination of Free Caustic Alkali

1. Weigh 20 grms of the sample into a 100 c.c. flask, dilute with approximately 50 c.c. of freshly boiled distilled water, add an excess of neutral barium chloride solution, 1 c.c. of phenol phthalein solution, make up to the mark and mix.
2. Allow the precipitate to settle, draw off 50 c.c. of the clear liquid and titrate with normal acid (N/1).
3. Calculate to percentage of Na_2O existing as caustic alkali.

II. Determination of Ash and Total Alkalinity

1. Weigh 2 to 5 grms of the sample in a platinum dish, burn off the glycerin over a luminous Argand burner or other source of heat giving a low flame temperature, the temperature being kept low to avoid volatilisation and the formation of sulphides.
2. When the mass is charred to the point that water will not become coloured by soluble organic matter, lixiviate with hot distilled water, filter, wash and ignite the residue in the platinum dish.
3. Return the filtrate and washings to the dish, evaporate and carefully ignite without fusion. Weigh the ash.
4. Dissolve the ash in distilled water and titrate total alkalinity, using as an indicator methyl orange cold or litmus boiling.

III. Determination of Alkali present as Carbonate

1. Take 10 grms of the sample, dilute with 50 c.c. distilled water, add sufficient N/1 acid to neutralise the total alkali found at II 4 (above), boil under reflux condenser for 15 to 20 minutes.
2. Wash down the condenser tube with distilled water, free from carbon dioxide, and titrate back with N/1 NaOH, using phenol phthalein as an indicator.
3. Calculate the percentage of Na_2O. Deduct the Na_2O found in I 3 above. The difference is the percentage of Na_2O existing as carbonate.

IV. Alkali combined with Organic Acids

The sum of the percentages of Na_2O found at I and III deducted from the percentage found at II is a measure of the Na_2O or other alkali combined with organic acids.

V. Determination of Acidity

Take 10 grms of the sample, dilute with 50 c.c. distilled water, free from carbon dioxide, and titrate with N/1 NaOH and phenol phthalein. Express in terms of Na_2O required to neutralise.

VI. Determination of Total Residue at 160° C.

For this determination the crude glycerin should be slightly alkaline with Na_2CO_3, not exceeding the equivalent of $0.2°/_{\circ}$ Na_2O, in order to prevent loss of organic acids. To avoid formation of polyglycerols this alkalinity must not be exceeded.

(*a*) PREPARATION OF GLYCERIN

1. Weigh 10 grms of the sample into a 100 c.c. flask, dilute with water and add the calculated quantity of N/1 HCl or Na_2CO_3 to give the required degree of alkalinity.
2. Fill the flask to 100 c.c., mix the contents and measure 10 c.c. into a weighed petri or similar dish 2·5 in. diameter and 0·5 in. deep, which should have a flat bottom.

    In the case of crude glycerins abnormally high in organic residue a less quantity is to be evaporated, so that the weight of organic residue does not materially exceed 30 to 40 milligrams.

(*b*) EVAPORATION OF THE GLYCERIN

1. Place the dish on a water-bath (the top of the 160° oven acts equally well) until most of the water has evaporated. From this point the evaporation is effected in the oven, the temperature of which is carefully maintained at 160° C. Approximately the correct results are obtained in an oven measuring 12 in. cube, having an iron plate ¾ in. thick lying on the bottom to distribute the heat. Strips of asbestos mill-board are placed on a shelf half-way up the oven. On these strips the dish containing the glycerin is placed.

If the temperature of the oven has been adjusted to 160° C. with the door closed, a temperature of 130° to 140° can be readily maintained with the door partially open, and the glycerin, or most of it, should be evaporated off at this temperature. When only a slight vapour is seen to come off, remove the dish and allow to cool.

2. Add 0·5 to 1 c.c. of water, and by a rotary motion bring the residue wholly or nearly into solution. Allow the dish to remain on a water-bath or top of the oven until the excess water has evaporated and the residue is in such a condition that on returning to the oven at 160° C. it will not spurt.

> The time taken up to this point cannot be given definitely, nor is it important. Usually two or three hours are required. From this point, however, the schedule of time must be strictly adhered to.

3. Allow the dish to remain in the oven at 160° C. for one hour, cool, treat the residue with water, and evaporate the water as before.

4. Bake the residue again for one hour, after which cool the dish in a desiccator over sulphuric acid and weigh.

5. The treatment with water, etc., is repeated until a constant loss of 1 to 1·5 mgms per hour is obtained.

Corrections to be applied to the Weight of the Total Residue

In the case of the acid glycerin a correction must be made for the alkali added. 1 c.c. N/1 alkali represents an addition of 0·022 grm. In the case of the alkaline crudes a correction should be made for the acid added. Deduct the increase in weight due to the conversion of the NaOH and Na_2CO_3 to NaCl. The corrected weight, multiplied by 100, gives the percentage of *total residue at* 160° C.

Preserve the total residue for the determination of the non-volatile acetylisable impurities.

VII. Organic Residue

Subtract the ash from the total residue at 160° C. Report as organic residue at 160° C. (*Note.*—It should be noted that the alkaline salts of organic acids are converted to carbonates on ignition and that the CO_3 radicle thus derived is not included in the organic residue.)

VIII. Moisture

This test is based on the fact that glycerin can be completely freed from water by allowing it to stand *in vacuo* over sulphuric acid or phosphoric anhydride.

1. Place 2 to 3 grms of very pure bulky asbestos, freed from acid soluble material (which has been previously dried in a water oven), in a small stoppered weighing bottle of about 15 c.c. capacity.

2. Keep the weighing bottle in a vacuum desiccator furnished with a supply of concentrated sulphuric acid, under a pressure equivalent to 1 to 2 mm. of mercury, until constant in weight.

3. Drop 1 to 1·5 grms of the sample carefully on the asbestos in such a way that it will all be absorbed. Take the weight again and replace the bottle in the desiccator under 1 to 2 mm. pressure until constant in weight.

> At 15° C. the weight is constant in about 48 hours. At lower temperature, the test is prolonged.
>
> The sulphuric acid in the desiccator must be frequently renewed.

ACETIN PROCESS

Reagents required

A. BEST ACETIC ANHYDRIDE. This should be carefully selected. A good sample should not require more than 0·1 to 0·2 c.c. normal NaOH when a blank is run on 7·5 c.c. Only a slight colour should develop during digestion of the blank.

B. PURE FUSED SODIUM ACETATE. The purchased salt is again completely fused in a platinum, silica or nickel dish, avoiding charring, powdered quickly and kept in a stoppered bottle or desiccator. It is most important that the sodium acetate be anhydrous.

C. A SOLUTION OF CAUSTIC SODA FOR NEUTRALISING, OF ABOUT N/1 STRENGTH, FREE FROM CARBONATE. This can be readily made by dissolving pure sodium hydroxide in its own weight of water (preferably water free from carbon dioxide) and allowing to settle until clear, or filtering through an asbestos or paper filter. The clear solution is diluted with water free from carbon dioxide to the strength required.

D. N/1 CAUSTIC SODA, FREE FROM CARBONATE. Prepared as above and carefully standardised. Some caustic soda solutions show a marked diminution in strength after being boiled ; such solutions should be rejected.

E. N/1 ACID. Carefully standardised.

F. PHENOL PHTHALEIN SOLUTION. Dissolve 0·5 per cent. phenol phthalein in alcohol and neutralise.

Procedure

1. Into a narrow-mouthed flask (preferably round-bottomed), capacity about 120 c.c., which has been thoroughly cleaned and dried, weigh accurately and as rapidly as possible 1·25 to 1·5 grms of the glycerin.

> The authors prefer to use a weighing bottle with about 10 c.c. glycerin. Weigh, pour out approximate amount, stopper again and re-weigh. Subtract weights = glycerin taken.

2. Add about 3 grms of the anhydrous sodium acetate, then 7·5 c.c. of

the acetic anhydride, and connect the flask with an upright Liebig condenser.

For convenience the inner tube of this condenser should not be over 50 cm. long and 9 to 10 mm. inside.

The flask is connected to the condenser by either a ground glass joint (preferably) or a rubber stopper. If a rubber stopper is used it should have had a preliminary treatment with hot acetic anhydride vapour.

3. Heat the contents and keep just boiling for one hour, taking precautions to prevent the salts drying on the sides of the flask.

4. Allow the flask to cool somewhat, and through the condenser tube add 50 c.c. of the carbon dioxide-free distilled water heated to about 80° C., taking care that the flask is not loosened from the condenser.

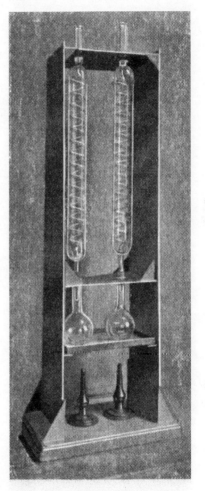

The object of cooling is to avoid any sudden rush of vapours from the flask on adding water, and to avoid breaking the flask. Time is saved by adding the water before the contents of the flask solidify, but the contents may be allowed to solidify and the test proceeded with the next day without detriment. The contents of the flask may be warmed to, but must not exceed, 80° C., until the solution is complete except a few dark flocks representing organic impurities

Fig. 59

in the crude. By giving the flask a rotary motion, solution is more quickly effected.

5. Cool flask and contents without loosening from condenser. When quite cold wash down the inside of the condenser tube, detach the

flask, wash off stopper or ground glass connection into the flask, and filter contents of the flask through an acid washed filter into a Jena glass flask of about 1 litre capacity.

6. Wash thoroughly with cold distilled water free from carbon dioxide. Add 2 c.c. of phenol phthalein solution (F), then run in a caustic soda solution (C) or (D) until a faint pinkish yellow colour appears throughout the solution.

    This neutralisation must be done *most carefully*; the alkali should be run down the sides of the flask, the contents of which are *kept rapidly swirling* with occasional agitation or change of motion until the solution is nearly neutralised, as indicated by the slower disappearance of the colour developed locally by the alkali running into the mixture. When this point is reached the sides of the flask are washed down with carbon-dioxide-free water and the alkali subsequently added drop by drop, mixing after each drop until the desired tint is obtained.

7. Now run in from a burette 50 c.c. or a calculated excess of N/1 NaOH (D) and note carefully the exact amount.

8. Boil gently for 15 minutes, the flask being fitted with a glass tube acting as a partial condenser.

9. Cool as quickly as possible and titrate excess of NaOH with N/1 acid (E) until the pinkish yellow, or chosen end point colour just remains.

    A further addition of the indicator at this point will cause a return of the pinkish colour; this must be neglected, and the first end point taken.

BLANK TEST

As the acetic anhydride and sodium acetate may contain impurities which affect the result it is necessary to make a blank test, using the same quantities of acetic anhydride and sodium acetate as in the analysis. After neutralising the acetic acid, it is not necessary to add more than 5 c.c. of the N/1 alkali (D) as that represents the excess of alkali usually left after saponification of the triacetin in the glycerol determination.

Calculation

Each c.c. N. NaOH absorbed = 0·030678 grm glycerol.

Example.   Bottle and glycerin (1) = 23·0275
                            (2) = 21·5715

weight of gly. taken   1·4560
c.c. N. HCl used for back titration =   8·25,
c.c. N. NaOH absorbed = 41·75,
blank absorbed =   0·25,
diff. = 41·5.

$$\text{Glycerin} = \frac{41 \cdot 5 \times \cdot 03067 \times 100}{1 \cdot 456}$$

= **87·42** per cent. glycerol.

Determination of the Glycerol Value of the Acetylisable Impurities

The total residue at 160° C. is dissolved in 1 or 2 c.c. of water, washed into a clean acetylising flask 120 c.c. capacity and the water evaporated. Now add anhydrous sodium acetate and proceed as in the glycerol determination before described. Calculate the result to glycerol.

Analysis of Acetic Anhydride

Into a weighed stoppered vessel, containing 10 to 20 c.c. of water, run about 2 c.c. of the anhydride, replace stopper and weigh ; allow to stand, with occasional shaking, for several hours, till all anhydride is hydrolysed ; then dilute to about 200 c.c., add phenol phthalein and titrate with $N/1$ NaOH. This gives the total acidity due to free acetic acid and acid formed from anhydride.

Into a stoppered weighing bottle, containing a known weight of recently-distilled aniline (from 10 to 20 c.c.), measure about 2 c.c. of the sample, stopper, mix, allow to cool, and weigh. Wash contents into about 200 c.c. cold water, and titrate acidity as before. This yields the acidity due to the original, preformed acetic acid plus one-half the acid due to anhydride (the other half having formed acetanilide); subtract the second result from the first (both calculated for 100 grms) and double result, obtaining c.c. $N/1$ NaOH per 100-grm sample. One c.c. NaOH equals 0·0510 anhydride.

DICHROMATE PROCESS

Reagents required

A. PURE POTASSIUM DICHROMATE powdered and dried in air free from dust or organic vapours, at 110° to 120° C. This is taken as the standard.

B. DILUTE SOLUTION OF DICHROMATE. Dissolve 3·7282 grms of (A) in distilled water and make up to 1 litre at 15·5° C.

C. FERROUS AMMONIUM SULPHATE. Dissolve 3·7282 grms of potassium dichromate (A) in 50 c.c. water, add 50 c.c. of 50 per cent. (by vol.) sulphuric acid, and to the cold, undiluted solution add from a weighing bottle a moderate excess of the ferrous ammonium sulphate, and titrate back with the dilute dichromate (B). Calculate the value of the ferrous salt in terms of dichromate.

D. SILVER CARBONATE. This is prepared as required for each test from 140 c.c. of 0·5 per cent. silver sulphate solution by precipitation with about 4·9 c.c. $N/1$ sodium carbonate solution (a little less than the calculated quantity of $N/1$ sodium carbonate should be used ; any excess of alkali carbonate prevents rapid settling). Settle, decant, and wash once by decantation.

E. SUBACETATE OF LEAD.   Boil a pure 10 per cent. solution of lead
   acetate with an excess of litharge for one hour, keeping the volume
   constant, and filter while hot.   Disregard any precipitate which
   subsequently forms.   Preserve out of contact with carbon dioxide.
F. POTASSIUM FERRICYANIDE.   A very dilute solution containing
   about 0·1 per cent.

Procedure

1. Weigh 20 grms of the glycerin, dilute to 250 c.c., and take 25 c.c.
2. Add the silver carbonate, allow to stand, with occasional agitation,
   for about 10 minutes, and add a slight excess (about 5 c.c. in most
   cases) of the basic lead acetate (E); allow to stand a further few
   minutes, dilute with distilled water to 100 c.c. and then add 0·15 c.c.
   to compensate for the volume of the precipitate.   Mix thoroughly.
3. Filter through a dry filter into a suitable narrow-mouthed vessel,
   rejecting the first 10 c.c., and return filtrate if not clear and bright.
   Test a portion of the filtrate with a little basic lead acetate, which
   should produce no further precipitate.   In the great majority of
   cases 5 c.c. is ample.

   > Occasionally a crude sample will be found requiring more, and in this
   > case another aliquot of 25 c.c. of the dilute glycerin should be taken,
   > and purified with 6 c.c. of the basic acetate.   Care must be taken to
   > avoid a marked excess of basic acetate.

4. Measure off 25 c.c. of the clear filtrate into a glass flask or beaker
   (previously cleaned with potassium dichromate and sulphuric acid).
   Add 12 drops of sulphuric acid (1 : 4) to precipitate the small excess
   of lead as sulphate.   Add 3·7282 grms of the powdered potassium
   dichromate (A).   Rinse down the dichromate with 25 c.c. of water,
   and stand, with occasional shaking, until all the dichromate is dissolved (no reduction will take place).
5. Now add 50 c.c. of 50 per cent. sulphuric acid (by volume) and
   immerse the vessel in boiling water for two hours, and keep protected
   from dust and organic vapours, such as alcohol, till the titration is
   completed.
6. Add from a weighing bottle a slight excess of the ferrous ammonium
   sulphate (C), making spot tests on a porcelain plate with the potassium ferricyanide (F).   Titrate back with the dilute dichromate.
7. From the amount of dichromate reduced calculate the percentage
   of glycerol.

   $$1 \text{ grm glycerol} = 7\cdot4564 \text{ grms dichromate,}$$
   $$1 \text{ grm dichromate} = 0\cdot13411 \text{ grm glycerol.}$$

Remarks

It is important that the concentration of acid in the oxidation mixture
and the time of oxidation be strictly adhered to.

Before the dichromate is added to the glycerin solution, it is essential that a slight excess of lead be precipitated with the sulphuric acid, as stipulated.

For crudes practically free from chlorides, the quantity of silver carbonate may be reduced to $\frac{1}{5}$ and the basic lead acetate to 0·5 c.c.

It is sometimes advisable to add a little potassium sulphate to ensure a clear filtrate.

Instructions for calculating actual Glycerol Content

1. Determine the apparent percentage of glycerol in the sample by the acetin process, as described. The result will include acetylisable impurities, if any be present.
2. Determine the total residue at 160° C.
3. Determine the acetin value of the residue (2) in terms of glycerol.
4. Deduct the result found at (3) from the percentage obtained at (1) and report this corrected figure as glycerol. If volatile acetylisable impurities are present, these are included in this figure.

Notes and Recommendations

Experience has shown that in crude glycerin of good commercial quality, the sum of water, total residue at 160° C. and corrected acetin result, come to within 0·5 of 100. Further, in such crudes, the dichromate result agrees with the uncorrected acetin result to within 1 °/₀.

In the event of greater differences being found, impurities such as *polyglycerols* or *trimethylene glycol* are present. Trimethylene glycol is more volatile than glycerin; it can therefore be concentrated by fractional distillation. An approximation to the quantity can be obtained from the spread between the acetin and dichromate results of such distillates, trimethylene glycol showing by the former method 80·69 °/₀ and by the latter 138·3 °/₀ expressed as glycerol.

If the non-volatile organic residue at 160° C. in the case of a soap lye crude be over 2·5 °/₀, i.e. when not corrected for carbon dioxide in the ash, then the residue shall be examined by the acetin method, and any excess of glycerol found over 0·5 °/₀ shall be deducted from the acetin figure.

In the case of saponification, distillation, and other glycerins, the limit of organic residue which should be passed without further examination shall be fixed at 1 °/₀. In the event of the sample containing more than 1 °/₀ the organic residue must be acetylated, and any glycerol found (after making the deduction of 0·5 °/₀) shall be deducted from the percentage of glycerol found by the acetin test.

BRITISH STANDARD SPECIFICATIONS FOR SOAP LYE AND SAPONIFICATION CRUDE GLYCERINS

Soap Lye Crude Glycerin

Analysis to be made in accordance with the International Standard Methods (I.S.M., 1911).

GLYCEROL. The standard shall be 80 °/₀ glycerol. Any crude glycerin tendered which tests 81 °/₀ glycerol or over, shall be paid for at a *pro rata* increase, calculated as from the standard of 80 °/₀. Any crude glycerin which tests under 80 °/₀ but is 78 °/₀ or over, shall be subject to a reduction of 1½ times the shortage, calculated at *pro rata* price as from 80 °/₀. If the test falls below 78 °/₀, the buyer shall have the right of rejection.

Thus :

Any premium to be paid would be based upon the price per unit of glycerol, e.g. an excess of 1·0 °/₀ in the glycerol content above the standard of 80·0 °/₀ would mean an additional $\frac{1}{80}$ on the contract price, and a deficiency of 1·0 °/₀ in the glycerol content would mean a deduction $\frac{1·5}{80}$ on the contract price.

Example :

Let X = corrected price.

Y = glycerol °/₀ by analysis (I.S.M.), say 81·20 °/₀.

Z = contract price, say £60 per ton.

S = standard °/₀ glycerol—80 °/₀.

Then $X = \dfrac{Z \times Y}{S}$.

Therefore $X = \dfrac{60 \times 81·2}{80} = £60. 18s. 0d.$ per ton.

Again, if $Y = 78·8$ °/₀.

Then $80 - 78·8 = 1·20$, difference × 1½ = 1·80.

And $80 - 1·80 = 78·2$.

Therefore $X = \dfrac{60 \times 78·2}{80} = £58. 13s. 0d.$ per ton.

ASH. The standard shall be 10 °/₀. In the event of the percentage of ash exceeding 10 °/₀, but not exceeding 10·5 °/₀, a percentage deduction shall be made for the excess calculated as from 10 °/₀ at contract price, and if the percentage of ash exceeds 10·5 °/₀ an additional percentage deduction shall be made equal to double the amount in excess of 10·5 °/₀. If the amount of ash exceeds 11 °/₀, the buyer shall have the right to reject the parcel.

Thus :

If the ash as shown by analysis (I.S.M.) does not exceed 10·5 °/₀, the excess over the standard (10·0 °/₀) is taken as a percentage on the contract price. In the same way any excess over

10·5 is multiplied by 2, calculated in the same manner and added to the deduction found for the excess up to 10·5 $\%$.

Example :

Let D = deduction from contract price due to excess over the standard (10·0 $\%$).

Z = contract price—say £60 per ton.

e = excess of ash up to 10·5 $\%$.

e' = excess of ash over 10·5 $\%$.

Then $D = \dfrac{Z \times e}{100} + \dfrac{Z \times 2e'}{100}$.

Therefore if ash is, say 10·4 $\%$,

$$D = \frac{60 \times 0·4}{100} = 4s.\ 10d.\ \text{per ton.}$$

And if ash is, say 10·8 $\%$,

$$D = \frac{60 \times 0·5}{100} + \frac{60 \times 0·3 \times 2}{100} = 13s.\ 2d.\ \text{per ton.}$$

ORGANIC RESIDUE.   The standard shall be 3 $\%$. A percentage deduction shall be made of three times the amount in excess of the standard of 3·0 $\%$ calculated at contract price. The buyer shall have the right to reject any parcel which tests over 3·75 $\%$.

Thus:

If the organic residue as shown by analysis (I.S.M.) exceeds 3·0 $\%$, the excess over this standard is multiplied by 3 and taken as a percentage on the contract price.

Example:

Let D = deduction due to excess over the standard (3·0 $\%$).

Z = contract price, say £60 per ton.

e = excess organic residue over the standard (3·0 $\%$).

Then $D = \dfrac{Z \times 3e}{100}$.

Therefore if analysis shows 3·25 $\%$ organic residue,

$$D = \frac{60 \times 0·25 \times 3}{100} = 9s.\ \text{per ton.}$$

Saponification Crude Glycerin

Analysis to be made in accordance with the International Standard Methods.

GLYCEROL.   The standard shall be 88 $\%$. Any crude glycerin tendered which tests 89 $\%$ or over, shall be paid for at a *pro rata* increase, calculated as from the standard of 86 $\%$. Any crude glycerin which tests under 88 $\%$, but is 86 $\%$ or over, shall be subject to a reduction of 1½ times the shortage, calculated at a *pro rata* price as from 88 $\%$. If the test falls below 86 $\%$, the buyer shall have the right of rejection.

Thus:

Any premium to be paid would be based upon the price per unit of glycerol, e.g. an excess of 1·0 °/$_o$ in the glycerol content above the standard of 88 °/$_o$ would mean an additional $\frac{1}{88}$ on the contract price, and a deficiency of 1·0 °/$_o$ in the glycerol content would mean a deduction of $\frac{1·5}{88}$ on the contract price. [For examples of calculations see "Soap crudes," p. 194.]

ASH. The standard shall be 0·5 °/$_o$. In the event of the percentage of ash exceeding 0·5 °/$_o$, a percentage deduction shall be made equal to double the amount in excess of the standard of 0·5 °/$_o$ calculated at contract price. If the amount of ash exceeds 2·0 °/$_o$, the buyer shall have the right to reject the parcel.

Example:

Let D = deduction from contract price due to excess over the standard (0·5 °/$_o$).

Z = contract price, say £68 per ton.

e = excess of ash over the standard (0·5 °/$_o$).

Then $D = \dfrac{Z \times 2e}{100}$.

Therefore, if analysis shows 1·20 °/$_o$ ash,

$$D = \frac{68 \times 0·7 \times 2}{100} = 19s. \text{ per ton.}$$

ORGANIC RESIDUE. The standard shall be 1 °/$_o$. A percentage deduction shall be made of twice the amount in excess of the standard of 1 °/$_o$ calculated at the contract price. The buyer shall have the right to reject any parcel which tests over 2 °/$_o$.

The example and formula for ash as above is equally applicable to organic residue.

SECTION VI

TESTING AND ANALYSIS OF HYDROCARBON OILS
AND WAXES

SECTION VI

HYDROCARBON OILS AND WAXES

STANDARD DETERMINATIONS

The standard **physical and chemical determinations** for the fatty oils and waxes are not all applicable to hydrocarbon (mineral) oils. The following yield useful information in many cases :

- (1) S. G. at 15·5° C.
- (2) M. P. ° C. (mineral waxes).
- (4) Refractive Index.
- (5) Viscosity.
- (6) Solubility.
- (7) Optical Activity.
- (8) Iodine Absorption.
- (10) Saponification Value.
- (14) Acid Value.

In addition, there are the following important general tests applying to mineral oils :

Flash point.
Ignition test.
Cold test.

There are also special tests for particular classes of oils, relating mainly to their industrial use.

For analytical purposes the hydrocarbon oils and waxes may be sub-divided as follows :

CRUDES
$\begin{cases} \text{Ozokerit} \\ \text{Petroleum} \\ \text{Bitumen} \\ \text{Coals and Shales} \end{cases}$

DISTILLED AND REFINED PRODUCTS
$\begin{cases} \text{Naphthas and " Spirits"} \\ \text{Illuminating oils} \\ \text{Lubricating oils} \\ \text{Fuel oils} \\ \text{Waxes} \end{cases}$

Specific Gravity ①

The best method for crudes is usually the s.g. bottle. If the oil is very viscous, the bottle may be filled hot, cooled to the normal tempera-ture (the shrinkage being made up as it cools), and the stopper inserted by gradually forcing home. Solids, as waxes, etc., may be ascertained by the flotation method (page 34).

Naphthas and illuminating oils may conveniently be tested by the hydrometer : for the heavier distillates the bottle is suitable.

Melting Point ②

The crude solid products are best tested by the platinum wire loop method (p. 41), as they do not give clearly defined melting points. The refined waxes may be tested by the capillary-tube method.

Refractive Index ④

Since most of the readings fall outside the scale of the butyro-refracto-meter, either the Abbé type or the Pulfrich may be employed. The former is very convenient, as it requires only a drop of the substance for the examination, and it has a range of from 1·3 to 1·7.

The determination of the heavier oils and waxes is best made at the standard temperature 40° C. (or calculated to this figure), but for the lighter distillates the temperature of 20° C. is more suitable on account of their volatility.

Viscosity ⑤

The viscosity determination is used practically exclusively for lubri-cating oils. It is best in all cases to standardise the efflux viscometer employed so that the readings may be converted into absolute terms. The "Viscosity ratio number" gives a measure of the efficiency of a lubricant at higher temperatures.

Solubility ⑥

This test has a special application in the case of coal-tar oils, which are soluble in dimethyl sulphate, while oils from other sources remain undissolved.

> The test is due to *Valenta* and is made as follows:
> In a 10 c.c. graduated cylinder place 2 c.c. of oil and 4 c.c. of dimethyl sulphate (very poisonous), shake for 1 minute and note difference in volume after a clean separation into two layers has taken place. The diminution is taken as due to coal-tar hydrocarbons. In the case of very volatile distillates, the method is not quite accurate.

Valenta also distinguishes between rosin oil and mineral oil by means of the solubility of the former in glacial acetic acid. Proceed as follows:

> Mix 2 c.c. of the oil in a test-tube with 10 c.c. of glacial acetic acid: immerse tube in water at 50° C. and keep agitated. Filter through a wet filter, and reject the first 4 c.c., collecting the next 4 c.c. and titrating a weighed quantity with standard alkali. Calculate the acetic acid present; the difference in strength is due to the oil dissolved. Rosin acids present will affect slightly the accuracy of the result, but this does not affect its general utility.

Solubility in alcohol is sometimes employed as a test for hydrocarbon oils. Thus kerosene is soluble in about 2 vols. of absolute alcohol.

Optical Activity ⑦

The determination of the optical activity of mineral oils is of limited use in the case of mixtures (or suspected mixtures) of rosin oils or of castor oil with mineral hydrocarbon oils, the latter having rarely a higher rotation than 1·2°. The test is carried out in the polarimeter exactly as for fatty oils.

Iodine Absorption ⑧

The Wijs-Hübl iodine solution is suitable, and the test is carried out exactly as for fatty oils, but it is often necessary to employ a measured excess of solution and to make a blank test side by side with a known specimen to eliminate the time factor, as in many cases definite absorption values are not obtainable, but the figure is largely proportional to the time of action.

Saponification Value ⑩

This is carried out in the same manner as described for fatty oils. Crudes and bituminous products may give positive figures owing to the esters of fatty acids (e.g. montanic) which they often contain. Refined oils give no saponification values, so that admixtures of fatty oils may readily be detected in this way. The mineral oil may float on the surface of the alcohol at the end of the boiling, but this does not affect the accuracy of the titration.

In the case of dark crudes or waxes, the colour change may be difficult to see. In this case the liquid, previous to titration, may be poured into a large bulk of hot distilled water, and a larger amount of indicator employed ; or spot tests with phenol phthalein may be made on a porcelain tile.

Acid Value ⑭

This test may be employed for the detection of rosin, fatty acids or mineral acids. In the latter case it is preferable to shake a given weight of the oil with hot distilled water in a separating funnel ; run off water, and titrate with alkali, using methyl orange as indicator.

FLASH POINT

Definition

The flash point of an oil is the temperature at which the oil gives off a vapour, inflammable or explosive in air.

Remarks

All the fatty oils have very high flash points, and do not evolve inflammable vapours until decomposition occurs. Mineral oils are very variable, the more volatile oils flashing at atmospheric temperatures. Since a low flash point renders illuminating oils unsafe, the test is of great importance in the examination of kerosene, the flash point of which is legally fixed at a minimum of 73° F. The standard test is made in the ABEL TESTER (*v. infra*).

In the case of lubricating oils, and heavier fractions, the GRAY or PENSKY-MARTENS' instrument is suitable.

The flash point of an oil is not necessarily determined by its volatility or mean boiling point, since a comparatively heavy oil may contain a small percentage of volatile constituents. When this occurs, it may be generally attributed to faulty methods of refinement.

The older method of making the test was in an open vessel, and is now known as the "open flash test." This is now seldom employed, a closed chamber being used for heating the oil, and an opening momentarily made by means of a mechanical device for the application of the flame used for testing.

THE ABEL TEST

The Apparatus

This consists of a metal vessel for the oil, which is placed in a hot-air chamber, heated by the water-bath, which is fitted with a funnel, an overflow pipe and a thermometer. The cover for the oil chamber has a perforated slide with holes corresponding with similar openings on the cover itself, and given positions of the slide open and close the chamber.

A small burner, fixed on a pivot, and used with wick or gas flame is made to dip at frequent intervals of time (every 4 seconds) into one of the openings, the action of opening also applying the flame.

The simpler types of apparatus are operated by hand, and timed by means of a pendulum supplied with the apparatus. The standard instrument is operated automatically by clockwork.

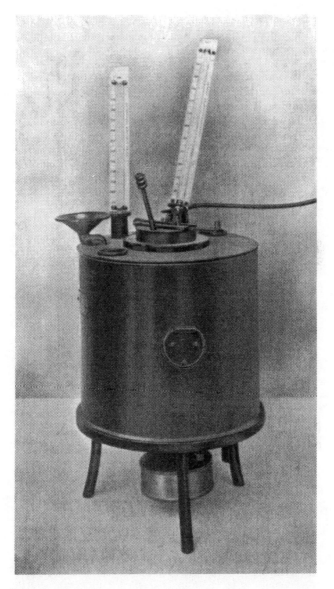

Fig. 60

Preliminary Operations

    (*a*) FILLING OIL CUP.

Place the oil cup (after removing cover) on a level surface in a good light, and fill with the oil to be tested, without splashing, until the surface of the oil is exactly level with the point of the gauge fixed in the cup.

Fill from a small vessel to avoid splashing.

Place the cover carefully on the cup, and press down into position, insert the *round bulb* thermometer in the socket provided, pressing carefully home, the scale to be facing the operator.

    (*b*) THE TEST-FLAME.

If oil is employed for the test-flame, moisten the cotton wool in the lamp and the wick with paraffin oil, light the wick, and adjust so that the flame is of the size of the bead fixed on the cover of the oil cup. If gas is used, connect up with rubber tube to gas supply, and adjust flame as stated.

    (*c*) TIMING.

If the simpler type is used, place the pendulum in position, suspended from the support supplied with the instrument. (The pendulum consists of a brass ball attached to a silk cord 24 inches in length from its point of suspension to the centre of gravity of the brass ball.)

    (*d*) PREPARING THE WATER-BATH.

Heat to boiling in a large beaker or basin about $\frac{1}{2}$ litre of water, and add sufficient cold water to reduce the temperature to 135° F. Pour this into the instrument through the funnel provided until it flows out of the overflow pipe. Insert the thermometer (long bulb) and adjust temperature to exactly 130° F. (by allowing to cool, or heating with spirit-lamp provided) before carrying out test.

Procedure

1. Having adjusted the temperature of the water-bath to exactly 130° F., place the oil cup and cover carefully in position. Adjust flame of test-burner if necessary, set pendulum swinging and observe thermometer in oil cup. (If clockwork type, wind up by turning knob from left to right.)

2. When temperature reaches 66° F., draw open slide, which motion applies test-flame to orifice.

    The opening of the slide should be done slowly, during the time of *three swings* (backwards, forwards, backwards) of the pendulum, and the slide quickly *shut* during the fourth swing of the pendulum. This is necessary to ensure uniformity in admission of air, and application of flame.

    (In case of automatic type, depress the trigger provided.)

3. If a flash occurs at this temperature (66° F.) (or at any point below 73° F.) the operation must be repeated with a fresh portion of oil cooled down to 55° F. before being placed in the cup ; and the first application of the test-flame made when the temperature reaches 60° F.

Fig. 61

Table for correction of Flash Points indicated by the Test for Variations in Barometric Pressure on either side of Thirty Inches.

Barometer in inches.

Flash Point in degrees Fahrenheit.

27	27·2	27·4	27·6	27·8	28	28·2	28·4	28·6	28·8	29	29·2	29·4	29·6	29·8	30	30·2	30·4	30·6	30·8	31
60·2	60·5	60·8	61·2	61·5	61·8	62·1	62·4	62·8	63·1	63·4	63·7	64	64·4	64·7	65	65·3	65·6	66	66·3	66·6
61·2	61·5	61·8	62·2	62·5	62·8	63·1	63·4	63·8	64·1	64·4	64·7	65	65·4	65·7	66	66·3	66·6	67	67·3	67·6
62·2	62·5	62·8	63·2	63·5	63·8	64·1	64·4	64·8	65·1	65·4	65·7	66	66·4	66·7	67	67·3	67·6	68	68·3	68·6
63·2	63·5	63·8	64·2	64·5	64·8	65·1	65·4	65·8	66·1	66·4	66·7	67	67·4	67·7	68	68·3	68·6	69	69·3	69·6
64·2	64·5	64·8	65·2	65·5	65·8	66·1	66·4	66·8	67·1	67·4	67·7	68	68·4	68·7	69	69·3	69·6	70	70·3	70·6
65·2	65·5	65·8	66·2	66·5	66·8	67·1	67·4	67·8	68·1	68·4	68·7	69	69·4	69·7	70	70·3	70·6	71	71·3	71·6
66·2	66·5	66·8	67·2	67·5	67·8	68·1	68·4	68·8	69·1	69·4	69·7	70	70·4	70·7	71	71·3	71·6	72	72·3	72·6
67·2	67·5	67·8	68·2	68·5	68·8	69·1	69·4	69·8	70·1	70·4	70·7	71	71·4	71·7	72	72·3	72·6	73	73·3	73·6
68·2	68·5	68·8	69·2	69·5	69·8	70·1	70·4	70·8	71·1	71·4	71·7	72	72·4	72·7	73	73·3	73·6	74	74·3	74·6
69·2	69·5	69·8	70·2	70·5	70·8	71·1	71·4	71·8	72·1	72·4	72·7	73	73·4	73·7	74	74·3	74·6	75	75·3	75·6
70·2	70·5	70·8	71·2	71·5	71·8	72·1	72·4	72·8	73·1	73·4	73·7	74	74·4	74·7	75	75·3	75·6	76	76·3	76·6
71·2	71·5	71·8	72·2	72·5	72·8	73·1	73·4	73·8	74·1	74·4	74·7	75	75·4	75·7	76	76·3	76·6	77	77·3	77·6
72·2	72·5	72·8	73·2	73·5	73·8	74·1	74·4	74·8	75·1	75·4	75·7	76	76·4	76·7	77	77·3	77·6	78	78·3	78·6
73·2	73·5	73·8	74·2	74·5	74·8	75·1	75·4	75·8	76·1	76·4	76·7	77	77·4	77·7	78	78·3	78·6	79	79·3	79·6
74·2	74·5	74·8	75·2	75·5	75·8	76·1	76·4	76·8	77·1	77·4	77·7	78	78·4	78·7	79	79·3	79·6	80	80·3	80·6
75·2	75·5	75·8	76·2	76·5	76·8	77·1	77·4	77·8	78·1	78·4	78·7	79	79·4	79·7	80	80·3	80·6	81	81·3	81·6

4. Observe thermometer carefully, and open slide in same manner at each degree rise until a flash occurs.

> Often, before the actual flash occurs, the test-flame will be seen to enlarge with a halo-like effect. The true flash is an instantaneous large blue flame which spreads over the whole surface of the oil, and usually extinguishes the test-flame.

5. Repeat with a fresh portion of oil.

6. Correction should be made for atmospheric pressure. The temperature of the flash varies 1·6° F. for each inch of the barometer. The barometer reading must therefore be taken at the time of the test, and correction applied by reference to the table on p. 206. The flash temperature is found in the column headed by the barometer reading in inches, and the figure taken in the same horizontal line in the standard column (30 inches).

> Thus, if the observed flash point = 70° F., and the barometer reading 28·8 inches, this corrected to 30 inches will be 72° F.

Students will kindly leave apparatus clean and empty after use.

PENSKY-MARTENS' APPARATUS

For flash point of lubricating oils, etc.

The Apparatus

This is a modified form of the Abel instrument (page 208) for the determination of the flash point of heavier oils, requiring a greater degree of heat than that attainable in the former tester.

The oil cup is provided with a closely fitting cover through the centre of which passes a shaft carrying the stirring arrangement, worked by means of the handle **d**. In another opening is fixed the thermometer **e**. The oil cup is placed in the metal heating vessel, and is enclosed by a jacket of hot air. The mantle **c** is to protect from loss of heat by radiation. The lid of the oil cup is perforated with several orifices which are opened and closed by means of coincident openings in a sliding cover, and is rotated by turning the spindle ending in the milled head **f**. This movement uncovers the holes and at the same instant tilts the test-flame on to the surface of the oil. The jet **g** is connected by means of the adapter and tubing supplied to a gas tap, and the size of flame adjusted by means of the screw-valve till very small. (A separate re-lighting jet is usually supplied in addition.)

A triple bunsen burner is supplied for heating the oil, and a piece of wire gauze, which may be turned aside if stronger heating is desired.

To remove the oil cup when hot, a special fork is provided.

Procedure

1. Fill the cup up to gauge with oil[1], replace cover, place cup in position, adjust flame, and light burner beneath.

2. Heat somewhat rapidly at first, until 30° C. below the expected temperature (make preliminary rough test if necessary), then at the rate of not over 4° C. rise per minute, agitating continuously

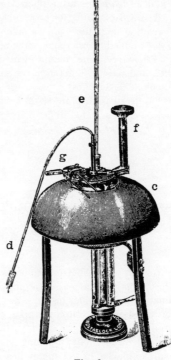

Fig. 62

by twisting the flexible connection **d** with the left hand. From time to time tilt the flame into the cup by turning the milled head **f** (with the right hand).

3. When the "flash" (a slight explosion, with large blue flame vanishing rapidly) occurs, note the temperature, and repeat with a fresh sample of oil, handling the oil cup with the fork-holder supplied.

Students will kindly leave cup and apparatus clean and dry after use.

[1] The oil must be free from moisture, as this interferes with the inflammability of the vapour produced.

GRAY'S FLASH POINT APPARATUS

The Apparatus

This is similar to the Pensky-Martens' instrument, but the oil cup shown in section, fig. 64, is fixed to the supports, and not removable except by unscrewing.

Like the Pensky-Martens' it has a hot-air jacket. Agitation of the liquid is provided for by stirrers, shown in detail in fig. 64. These are rotated by means of a steel shaft, bearing a small bevelled wheel, with toothed edge, geared with a vertical bevelled wheel and operated by means of the handle. Three openings are provided on the lid for the flash, and these are closed by a loose cover provided with openings which coincide with the ports in the fixed lid when the cover is turned one-quarter round, the movement causing the test-flame to dip into the cup.

This movement is caused by throwing the bevel gears out of play with a lateral movement of the horizontal shaft, causing the latter to engage with a pin operating the cover and jet. The cup is closed again by rapidly reversing the handle.

A thermometer and fitting are supplied with the instrument.

Procedure

1. Fill the cup—which

Fig. 63

must be clean and dry—with the sample up to the gauge (a line cut round the inside $1\frac{1}{2}$ inches from bottom). Replace the lid : insert thermometer and light the test-jet and adjust the flame to about $\frac{1}{8}$ inch long.

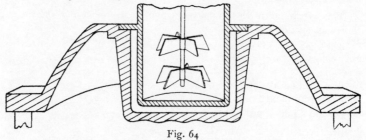

Fig. 64

2. Heat the oil cup with the bunsen burner whilst continuously rotating the stirrers, allowing the oil to rise about 20° C. per minute until near its expected flash point (make a rough test first if necessary) and afterwards at about one-fifth this rate.

3. Slide the gears out of position, open cup and apply test-jet by turning handle a quarter of a turn after its engagement with the cover-rotating device. Close by reversing the movement and re-engage the bevel wheels so as to keep oil in constant agitation.

    The whole movement is very easily and rapidly accomplished.

4. The temperature at which a slight explosion occurs is noted as the flash point of the oil.

Students will kindly leave cup and apparatus clean and dry after use.

IGNITION TEST

Definition

The ignition point of an oil is the lowest temperature at which the oil will continue to burn after application of a flame to its surface.

Requirements

A porcelain crucible $1\frac{1}{2}$ inches in height by $1\frac{1}{2}$ inches in diameter, embedded to about $\frac{3}{4}$ inch in sand on a sand-bath and tripod.

Small taper, or fine gas flame.

Thermometer reading to 350° C. or over.

Procedure

1. Fill crucible with the oil to be tested to about $\frac{1}{4}$ inch from its brim, and see that it is protected from draughts.

2. Stir gently with the thermometer and heat fairly rapidly at first until the temperature of the flash point of the oil; then turn burner down, and heat very slowly to about 15° C. above this temperature.

3. After every rise of 2° C. above the flash point apply the flame to the surface of the oil until the latter catches fire and burns quietly without further assistance.

COLD TEST. ("Setting point.")

Definition

The setting point of an oil is a little indefinite and is variously interpreted. Broadly it signifies the temperature at which the oil ceases to run.

Remarks

This test is greatly complicated by the fact that oils have often to be maintained at definite temperatures for fairly long periods before setting takes place, owing to the liability of *supercooling*. Further, oils kept in agitation give different results from those which are cooled without stirring.

Again, oils which have been previously heated give different results from normal oils.

It is necessary, therefore, to follow more or less arbitrary conditions in applying this test, and in many cases only an approximate figure can be obtained.

Oils differ considerably on being cooled according to the percentage of paraffin wax which they contain. The deposition of wax is accompanied by turbidity, but some oils set to a salve-like consistency while remaining quite clear.

Methods

THREE METHODS may be employed :

1. The oil may be cooled till wax separates, and warmed with stirring until all wax has melted, the setting point being taken as this temperature.

2. The oil may be cooled for several hours at definite temperatures, obtained by the use of freezing mixtures of varying compositions, and its behaviour noted. Setting point = temperature when oil ceases to run.

3. A definite degree of viscosity may be stipulated, and the temperature at which this is attained may be taken as the setting point.

   [These three methods are those specified by the following bodies :
   Method 1.   Scottish Mineral Oil Association.
     „   2.   New York Produce Exchange.
     „   3.   Prussian State Railways.]

1. Scotch method

Procedure

1. Take a short test-tube, about $1\frac{1}{4}$ inches in diameter, and run in the oil to about 2 inches depth.

2. Immerse in ice and salt freezing mixture, and slowly stir with a thermometer until well past the turbidity point.

3. Remove from ice, hold up to a source of light and stir with the thermometer until the last trace of wax has disappeared, taking the temperature at this point.

4. Repeat with the same sample of oil until two concordant results are obtained.

2. Freezing mixtures method[1]

This method depends upon the use of solutions of various salts in such proportions that they are saturated at their freezing points. In these conditions a constant temperature is maintained until the solutions are entirely frozen or reliquefied, and by the employment of various salts a range of low temperatures is made available.

Saline Freezing Solutions.

Salt	Parts per 100 of water	Freezing points of solutions	
Potassium sulphate	10	28·6° F.	− 1·9° C.
Sodium carbonate crystals	20	28·4	− 2·0
Potassium nitrate	13	26·9	− 2·85
Potassium nitrate } Sodium chloride }	13 } 3·3 }	23·0	− 5·0
Barium chloride	35·8	16·3	− 8·7
Potassium chloride	30	12·4	− 10·9
Ammonium chloride	25	4·3	− 15·4
Ammonium nitrate	45	1·85	− 16·75
Sodium nitrate	50	0·05	− 17·75
Sodium chloride	33	− 4·3	− 21·3

In testing a range of samples, a stand is employed which holds a number of test-tubes, and these are immersed in one of the above solutions contained in a vessel immersed in an ice and salt freezing mixture. As soon as this liquid commences to freeze (freezing is assisted by stirring, or adding a small quantity of frozen solution to the bulk) it will maintain its particular temperature constant as long as it remains only partially frozen, which can be attained by lifting it out of the freezing mixture occasionally. After a few hours the tubes are examined, and those oils which are set are noted, whilst the rest are tested in a liquid giving the next lowest temperature in the table, and proceeding in this manner until the temperature of − 21·3° C. has been attained, when, if the oil still remains liquid, it may be further cooled by means of liquefied gases. (This is however rarely necessary, except for use in special circumstances, e.g. aircraft, etc.) The various salt solutions are conveniently kept in stoppered bottles ready for use.

3. Fluidity method (Schultz)

This is a much more complicated method, and depends upon arbitrary conditions prescribed by the Prussian State Railway Direction. The apparatus employed is shown in the annexed diagram.

[1] Hofmeister, *Mitteilungen*, **7**, 24 (1889).

The Apparatus

a is a U-tube of 6 mm. internal diameter. This is held by a clip as shown, and is immersed in a cylindrical metal vessel **b** filled with a liquid freezing at the required temperature (either -5° C. or -15° C.) and surrounded by an ice and salt freezing mixture **c**. The tube is connected by the 3-way pieces as shown with the simple manometer **d** and the weighted funnel **e**, contained in the vessel **f**, and two clips are placed on the rubber tubes, one communicating with the air and the other with the funnel and manometer.

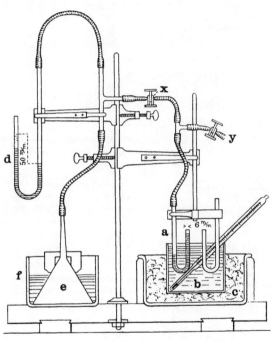

Fig. 65

Procedure

1. Introduce the sample of oil into the U-tube **a** by means of a pipette and immerse in the freezing fluid **b**, the level of the oil being 10 mm. below the fluid when the latter has obtained the correct temperature; allow the tube to remain immersed for 1 hour or more.

2. Pour water into vessel **f** until a pressure of 50 mm. is registered on the manometer **d**. Then open pinchcock **x**. The oil in the U-tube should rise *at the minimum rate of* 10 *mm. per minute.* If this is the case, the oil is considered to remain fluid at this temperature. "Winter oil" is tested at -15° C. and "summer oil" at -5° C.

3. The pressure is allowed to act on the oil for exactly 1 minute, and
the pinchcock **y** is then released, and the height of the oil measured
by the wet surface remaining when the oil has flowed back to level
in the **U**-tube.

A similar examination is made of oil heated to 50° C. for 10 minutes
and cooled for 1 hour.

EXAMINATION OF CRUDES

Crude Petroleum

SPECIFIC GRAVITY, see page 26.

DETERMINATION OF MOISTURE—the distillation method using a solvent
is most suitable (page 18).

SUSPENDED IMPURITIES, see page 19.

FLASH POINT. The Abel or Pensky apparatus is employed according
to the volatility of the oil (pages 207, 209).

Fractional Distillation (*Engler*)

This gives a measure of the yield of the various distillates from crudes,
and although, owing to the widely
differing conditions of the distillation, it does not give the same yields
which would be obtained on the
large scale, it gives good comparative indications provided the standard
conditions are always adhered to.

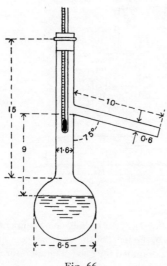

Fig. 66

The Engler flask employed has
the dimensions shown, and is used
with a condenser 60 cm. in length.
100 c.c. of oil are taken for the distillation, and the heat should be such
that 2 drops of condensate fall per
second. The fractions are collected
and measured between the following
temperatures :

Naphtha up to 150° C.
Illuminating oil 150--300° C.
Lubricating oils over 300° C.

The distillation is stopped when
the bottom becomes free of liquid,
or white vapours appear.

Formolite Reaction

This test is due to *Nastjukoff*[1]. It depends upon the fact that when

[1] *Petrol.* 1909, **4**, 1336.

formaldehyde, in the presence of sulphuric acid, is added to mineral oils the *unsaturated cyclic hydrocarbons* are precipitated, giving a yellow solid substance ("formolite"). The formolite number is the number of grms of dried formolite obtained from 100 c.c. of oil under the conditions of the test.

American oils give higher formolite values than Russian, owing to their content of unsaturated naphthenes. The amount of unsaturated hydrocarbon present is found by multiplying the formolite number by $\frac{4}{5}$.

Procedure (Marcusson's[1] modified method)

1. Dissolve 27 grms of the oil in 50 c.c. petroleum ether in a 300 c.c. flask, and add 30 c.c. conc. sulphuric acid (without shaking).
2. Cool with ice water, add 15 c.c. of 40 per cent. formalin, and shake flask in cooling mixture till no more heat is evolved.
3. Stand $\frac{1}{2}$ hour at room temperature and transfer to litre flask containing 200 c.c. of ice water, rinsing with water. Neutralise acid with excess of ammonia, and filter on a vacuum filter or Buchner funnel.
4. Wash with (1) petrol. ether, (2) water, and dry at 105° C. to constant weight.

    From the petrol. ether solution, the non-reacting portion of the oil can be obtained by evaporation.
5. The formolite number (grms per 100 c.c.) may be calculated from the above, knowing the S. G. of the oil.

Crude Ozokerite

Crude ozokerite is tested for its yield of ceresin or refined wax. The method is due to Von Boyen[2].

Heat 5 grms crude wax with 0·45 c.c. (15 per cent.) conc. H_2SO_4 to 180—200° till all odour of SO_2 disappears. Stir into hot mass 10 per cent. of animal charcoal (or ferrocyanide residues, dried at 140° C.) and about 6 grms of extracted sawdust. Cool, and treat in an extractor (Stokes', p. 22, is suitable) with petrol. ether for 2 hours. Remove, powder, re-extract. The combined extracts after evaporation of solvent give the yield of refined wax.

Crude Bitumen

The following test[3] is suitable :

Powder 2 grms of the material and treat with a saturated solution of ether in conc. hydrochloric acid, until the carbonate is completely decomposed, add 5 c.c. of ether to make up for evaporation. Digest for further 10 minutes, add 15 c.c. water and evaporate off ether. Filter, wash bitumen free from mineral acid and dry at 110° C. for 1 hour. Dissolve in chloroform, filter, wash residue, evaporate off solvent from filtrate, dry at 105° and weigh. The weight gives the amount of bituminous material present.

[1] *Chem. Zeit.* 1911, **38**, 729.
[2] *Zeit. f. angew. Chem.* 1898, **11**, 383.
[3] Prettner, *Chem. Zeit.* 1909, **33**, 917.

Crude Coals and Shales

The examination follows the lines of the commercial process, a trial destructive distillation is made on a small scale, and the yield of crude oil or tar, and sulphate of ammonia, gas, etc., calculated. The crude oil and tar can then be tested by fractionation, and the amounts of each distillate obtained noted (page 214).

EXAMINATION OF DISTILLATES

Naphtha

SPECIFIC GRAVITY, page 26.

REFRACTIVE INDEX, page 45.

SOLUBILITY, page 76.

FLASH POINT, page 202.

The naphtha may have to be cooled in a freezing mixture before testing.

FRACTIONAL DISTILLATION, page 214.

100 c.c. of the naphtha are distilled from the Engler flask and fractions are taken *every* 10° C. The temperature at which the first drop falls from the condenser is taken as the beginning of the boiling, and the end temperature is taken when the bottom of the flask becomes free from liquid.

Strictly the distillates should be taken at the figure at which water boils above 100° C. instead of the even 10°.

AROMATIC HYDROCARBONS.

Burton's test with fuming nitric acid (page 230) is recommended. The nitro-bodies are extracted with ether on dilution of the acid solution with water.

DETECTION OF TURPENTINE AND PINE OIL.

Add 1 drop of Wijs-Hübl iodine solution to several c.c. of the naphtha. Pure naphtha remains red in colour for at least 30 minutes without apparent change in intensity. In the presence of turpentine and pine oil the colour disappears or diminishes in intensity.

Determine amount present by iodine absorption value (page 92).

Illuminating Oils

(Normally fraction 150—300° C.)

COLOUR.

A good kerosene should be perfectly transparent in a 3-inch layer, and should have only a faint trace of yellow colour. *Colorimeters* are employed to determine the relative colour, on which kerosene is sold.

SPECIFIC GRAVITY, page 26.

COLD TEST, page 211.

Kerosene should remain liquid at − 10° C.

FLASH POINT, page 202.

The legal limit in this country is fixed at 73° F.; in Germany and the continent generally, at 21° C.

FRACTIONAL DISTILLATION, page 214.

As for naphtha (page 216), but fractions are conveniently taken distilling under 150° C. and every 25° C. up to 300° C., and the amount remaining in the still measured.

FORMOLITE TEST, page 214.

SOLUBILITY, page 76.

Kerosene is easily soluble in 2 vols. of absolute alcohol.

IODINE VALUE, page 92.

Russian kerosene o—1·6.

American   ,,   5·5—16·5.

Galician   ,,   o·1.

This distinguishes American kerosene from oils of other sources.

Lubricating Oils (mineral)

SPECIFIC GRAVITY, page 26.

The S.G. yields information as to the identity of lubricating oils, and gives indications of adulteration with rosin and tar oils.

VISCOSITY, page 62.

The most important test for this class of oil. The viscosity, though not a definite measure of lubricating value, is a reliable guide in many directions, since certain defined minimum viscosities are essential under the various conditions in which lubricants are employed.

The viscometer used should be standardised to give results in absolute measure, so as to be strictly comparable with results obtained with other forms of apparatus (see page 68). There is no reason why the figure ($\eta \times 100$) should not form a convenient expression of viscosity in all commercial transactions and specifications.

The "viscosity-ratio-number" (page 70) is rapidly ascertained, and gives useful information of the fall in viscosity at higher temperatures, and therefore under the actual working conditions.

IODINE ABSORPTION AND SAPONIFICATION VALUE, pages 92 and 106.

These are both important in the examination for rosin and fatty oils. The iodine values and saponification values of the various fatty oils, fats, and rosin are given in Vol. I. The iodine values of mineral lubricating oils are small, and the sapon. values nil.

ACID VALUE, page 123.

Useful for detecting the presence and amount of mineral acid (left from refining), page 123, and of fatty acids or rosin acids.

FLASH POINT, page 202.

COLD TEST, page 211.

A most important test, since, if the setting point is too high, the oil may be prevented from reaching the part to be lubricated.

MOISTURE, page 18.

GUMMING.

Mineral oils, if properly refined, are free from gumming tendencies.

(Of the fatty oils used as lubricants, only those with low iodine values are suitable.)

SUSPENDED MATTERS, page 19.

Graphite is often deliberately added to lubricants, and in many cases is a distinct advantage. Other suspended matters must be regarded as impurities.

ANTIFLUORESCENTS.

These are added to destroy or lessen the "bloom" (fluorescence) of lubricating oils. The chief substance used for this purpose is *nitro-naphthalene*. It may be detected as follows :

(1) Boil 2 c.c. of the sample for a few minutes with 4 c.c. of 2N alcoholic potash. In the presence of nitronaphthalene a blood-red to violet colour is produced by the reduction to azo-compounds.

(2) On treatment of the oil with zinc dust and dilute hydrochloric acid, a repulsive odour due to α-naphthylamine will be produced.

Scents

The most commonly employed substance to mask unpleasant odours in oils is *nitrobenzene*, easily recognised by its odour of bitter almonds

Coal-tar Oils

See *Valenta's* test with dimethyl sulphate (page 200).

Fuel Oils (Astatki, etc.)

The examination usually comprises :

MOISTURE DETERMINATION, page 18.

FRACTIONATION TEST (volatile components), page 214.

FLASH POINT, page 202.

COLD TEST, page 211.

WAXES

Paraffin Wax

MOISTURE AND MECHANICAL IMPURITIES, page 18.

YIELD OF PARAFFIN WAX (*Holde*).

Dissolve 0·5—1·0 grm in ether in a test-tube, avoiding excess, cool to - 21° C., and precipitate with an equal volume of absolute alcohol. If too thick to filter, add some alcohol-ether. Filter, dry and weigh. Evaporate the solvent, and repeat the operations with the residue, adding any wax resulting to the yield previously obtained.

ROSIN—acid value, etc.

MELTING POINT, page 38.

FATS—sapon. value and iodine value, pages 106 and 92.

Ozokerite (ceresin)

MELTING POINT, page 38.

ROSIN—acid value; shaking out, after saponification, with solvents (page 106) or extraction with 70 per cent. alcohol.

FATS—saponification value, and as for rosin.

PARAFFIN WAX.

The method was worked out by *Landsberger*:

1. Dissolve 1 grm in 30 c.c. chloroform by gentle heat in a 200 c.c. flask.
2. Cool to 20° C., add 60 c.c. of 96 per cent. alcohol, and stand for 20 minutes.
3. Collect ppt. on Buchner funnel (*a*).
4. Evaporate off solvent from filtrate, dissolve residue in 5 c.c. chloroform, cool to 20° C., add 15 c.c. 96 per cent. alcohol, and after standing 20 minutes, collect ppt. on Buchner funnel as before (*b*).
5. Dissolve (*a*) and (*b*) separately in hot benzene, evaporate and weigh. Weigh residue from filtrate after evaporation of solvent. Test on butyro-refractometer at 90° C.

Pure ceresin (ozokerite) with mixtures of paraffin (M.P. 53°) gave the following:

	ppt. *a*	ppt. *b*	residue
Pure ceresin ...	11·3—18·5	6·4—8·5	15·8—24·5
+ 10 °/₀ paraffin ...	10·6—19·5	3·5—9·0	8·5—14·5
+ 20 °/₀ ,, ...	11·0—12·9	2·5—2·8	7·2— 9·3

Montan Wax

MELTING POINT, page 38.

ACID VALUE, page 123.

Dissolve in mixture of ethyl and amyl alcohol and titrate with N/10 alcoholic potash [1].

SAPONIFICATION VALUE, page 106.

For complete saponification [1] it is best to employ 2 grms of wax, 25 c.c. of N/2 alcoholic (absolute) potash with 40 c.c. benzene. Titrate with N/2 HCl.

ADDED MATERIALS, as for ozokerite and paraffin.

[1] Marcusson, *Chem. Rev.* 1908, **15**, 193.

SECTION VII

THE TESTING AND ANALYSIS OF ROSIN AND
TURPENTINE

SECTION VII

THE TESTING AND ANALYSIS OF ROSIN AND TURPENTINE

ROSIN (Colophony)

A. DETECTION

(*a*) **In mineral oils**

As rosin has a high *acid value* its presence may be suspected in mineral oils showing distinct acidities. The acid value of the oil is determined (as on page 123), each per cent. of rosin present being approximately equivalent to 1·5 acid value.

Confirm as follows :

1. Take 10 c.c. of the oil, warm, and shake well with 10 c.c. of 70 per cent. alcohol : cool and filter through a moistened filter paper.
2. Evaporate filtrate : if rosin is present, residue should possess a gummy consistency. Test by the Liebermann-Storch test (below).

(*b*) **In fatty oils or fatty acids**

Rosin, in admixture with fatty oils or acids, is best detected by the **Liebermann-Storch test** as follows :

1. Gently warm a small quantity of the sample with an equal volume of pure acetic anhydride : cool.
2. Pipette off the acetic anhydride layer, and add to it one drop of sulphuric acid, S.G. 1·53 (by mixing 34·7 c.c. conc. H_2SO_4 with 35·7 c.c. water).
3. In presence of rosin a violet coloration (rapidly fugitive) is produced.

Remarks

i. If the sample is very dark in colour, and the reaction, for this reason, indistinct, saponify[1] with alcoholic potash, acidify and examine the fatty and rosin acids liberated.
ii. Cholesterol gives a similar coloration. It may be removed by dissolving out from the soap solution (i. above) with ether before decomposing with mineral acid.

[1] page 106.

(c) **In waxes**

If mineral waxes only are present, by the acidity (1·5 acid value = 1 per cent. rosin).

In beeswax and vegetable waxes, by the Liebermann-Storch test (above).

B. DETERMINATION

(a) **In mineral oils and waxes**

Approximately by acid value (above) or by saponifying the sample with alcoholic KOH, evaporating off alcohol, dissolving in water, shaking out the unsaponifiable matter with ether, and acidifying the aqueous liquor and filtering off and weighing the liberated rosin acids.

(b) **In fatty oils, acids, or natural waxes**

In oils and waxes the fatty and rosin acids are first obtained by saponification and liberation with mineral acids.

The best method of estimation in admixture with fatty acids is that proposed by **Twitchell**. This is based upon the formation of esters by fatty acids when acted upon by HCl gas in alcoholic solution, rosin acids remaining unaffected.

Twitchell Method

Apparatus and Reagents required

Apparatus fitted up as in photograph.

HCl is generated at **a** by dripping conc. sulphuric acid through dropping funnel into strong HCl solution, and the gas washed by the sulphuric acid washing bottle **b**.

It then passes through the bulb-tube **c** into the alcoholic solution of acids in flask **d** kept cool in running water (or constantly renewed cold water) in **e**.

Absolute alcohol (weaker alcohol is unsuitable).

Hydrochloric and sulphuric acids.

Ether.

N. alkali

Pumice.

The saturation with HCl should preferably be carried out in fume cupboard.

Procedure

1. Weigh accurately into a flask 2—3 grms of the mixed fatty and rosin acids, and dissolve in ten times the vol. of absolute alcohol.
2. Pass dry HCl gas through the solution, kept cool in circulating cold water, until no more gas is absorbed (about one hour). The gas is at first rapidly taken up. Allow to stand one hour.
3. Dilute contents with 150 c.c. water, add some powdered pumice and boil until the aqueous solution has become clear.

4. Transfer to separating funnel, cool to 15°, and rinse out flask with ether several times, adding this to separator. Insert stopper and shake vigorously. Settle: run off acid aqueous layer.

5. Wash ethereal solution of esters and rosin acids with water till the washings are no longer acid to litmus. Add 50 c.c. alcohol and run into titrating flask, rinsing separator with a little alcohol.

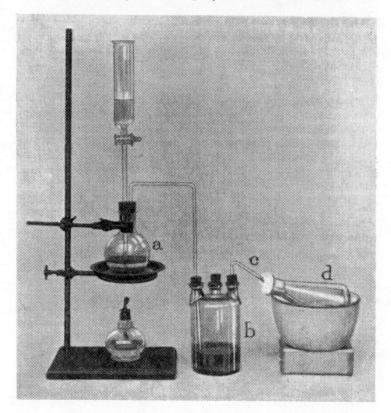

Fig. 67

6. Titrate with N. alkali, using phenol phthalein as indicator.

7. C.c. of N. alkali × 0·346 = rosin acids present.

The same method may be employed for estimating the rosin present in soaps, the mixed fatty acids being first isolated by acidifying and dried at 100° C.

TURPENTINE

Genuine turpentine is the distillation product of the "natural gum" of conifers. It is liable to adulteration by:

"Russian turpentine" | from distillation of roots and wood.
"Wood turpentine" |

Pine oil.
Rosin spirit.
Petroleum "naphthas" and "spirits."
Benzols.
Carbon tetrachloride, and other chlorinated hydrocarbons.

The standard determinations yielding information as to the purity of turpentine are:

Specific gravity at 15·5° C. (1)

Refractive index. (4)

Solubility. (6)

Optical activity. (7)

Iodine absorption. (8)

In addition, there are several specific tests of value for the detection of various adulterants.

Specific Gravity at 15·5° C. (1)

The S.G. is conveniently obtained by means of the S.G. bottle, but any of the standard methods may be employed.

The following table shows the S.G. of turpentine and some of its adulterants:

Name			S.G. at 15·5° C.
Pure American turpentine	...	...	0·865—0·870
Petroleum distillates			
B.P. 100—180° American	...	...	0·730—0·740
Russian	...	...	0·765—0·785
E. Indian	...	...	0·780—0·800
B.P. 180—200° ...	...	...	0·760—0·820
Coal-tar distillates			
Benzols	...	...	0·882—0·887
Toluol	...	...	0·868—0·872
Pine oil	...	...	0·854—0·874
Carbon tetrachloride	...	...	1·4

Refractive Index (4)

The refractive index of turpentine may be determined on the butyro-refractometer, as it comes within the scale readings of that instrument; but since some of the common adulterants of turpentine fall outside the scale, the Abbé type is the best for all-round work.

The refractive index is a useful figure in the examination of turpentine, especially when combined with the fractional distillation of the sample (*v. infra*). The temperature employed in the test is suitably lower than that for the fixed oils, on account of its volatility, say 20° C., the reading, if taken at a different temperature, being calculated to 20° C., using the correction ·00042 per degree.

The following table gives the usual variations of the refractive indices and scale readings of pure turpentine, and some of its substitutes:

	Refractive index 20° C.	Scale reading (z.b.) 20° C.
Genuine turpentine	1·469—1·472	64·8—69·5
Wood turpentine	1·466—1·468	60·1—63·2
Petroleum distillates		
Naphthas 100—180° B.P. ...	1·417—1·448	* —33·9
Coal-tar distillates		
Benzol	1·500	*
Toluol	1·487	95·5
Xylol	1·494	*
Solvent naphtha	1·496—1·499	*
Heavy benzol	1·523	*

* Outside limits of scale.

Solubility (6)

Turpentine is miscible with aniline, acetic anhydride, and dissolves in 10 parts of 90 per cent. alcohol. Most mineral oil distillates are insoluble in these solvents.

Frey's[1] Test

10 c.c. of turpentine is violently shaken in a stoppered 50 c.c. cylinder with 30 c.c. of anhydrous aniline for 5 minutes and is then allowed to stand. In the presence of petroleum distillates these will rise and float on the surface of the aniline, and their volume may be read off.

Marcille[2] proposes the following test, based upon its solubility in acetic acid:

Add 2 per cent. of water to glacial acetic acid. Test by treating the acid with an equal volume of dry CS_2 at 40° C.: a clear solution should result. Take 5 c.c. of turpentine, place in a graduated cylinder (stoppered) and gradually add the acetic acid in small amounts at a time, shaking vigorously, until a clear solution is obtained. Cool in water till turbid, and add small quantities of acid till a clear solution is again obtained which persists after cooling for 5 minutes.

The total volume of the liquid is then measured, the temperature taken, and

[1] *Jour. Amer. Chem. Soc.* 1908, **30**, 420.
[2] *Bull. Soc. Chim.* 1912, **11**, 762.

the ratio of solvent added to turpentine calculated, a correction of 0·08 being made for each 1° C. by which temperature differs from 15° C.

Pure turpentine dissolves in 3·7 parts of acid : petroleum spirit and rosin oil in 22 parts, and benzene in 0·25 part.

Optical Activity ⑦

Turpentine appears to vary very considerably in optical activity, observations having been made as divergent as from +18° to −26° (100 mm. tube at 20° C.).

Most of the usual adulterants of turpentine are optically inactive.

Iodine Absorption ⑧

The iodine absorption of turpentine varies to a considerable extent with the excess of iodine solution employed and the time of action. It is therefore advisable to perform the test side by side with a control specimen of pure turpentine. Under these conditions, valuable information is yielded by this determination.

The following results were obtained, employing an excess of 100 per cent. of iodine, and a time of action of 16 hours (Wijs' solution) :

Iodine absorption value, using 100 per cent.
excess iodine for 16 hours

American turpentine	400—410
Russian ,, ...	315—325
Wood ,, (very variable) ...	250—380
Rosin spirit	160—180
Pine oil	100—120
Petroleum distillates	very small

SPECIFIC TESTS FOR ADULTERANTS

A. PETROLEUM DISTILLATES

Fractionation Test

The best general test for ascertaining the purity of turpentine is to fractionally distil a given quantity and note the volume of distillate obtained between definite temperatures. In the case of suspected adulteration, the refractive index or the specific gravity of each fraction is taken, and a good indication of the extent of the adulteration is thus obtained.

The refractive index is the most rapidly obtained and suitable figure for this purpose, and the test has the great advantage of requiring only a drop of liquid for examination.

Procedure

1. Have a number of graduated cylinders ready, say 10 c.c. capacity, graduated in $\frac{1}{10}$ c.c.

2. Place in an Engler flask (page 214) 100 c.c. of the sample of turpentine, insert thermometer and cork, connect to condenser, place a cylinder to catch distillate, and heat with a moderate flame (keeping flame always the same height, and employing the same burner).

3. Note the temperature at which the *first drop* of liquid condenses, then observe thermometer closely throughout distillation. When the temperature reaches a round number of degrees, ending in 0 or 5, remove cylinder, and immediately replace with a fresh one without disturbing distillation. Note the number of c.c. of the first fraction, and ascertain refractive index (and) or specific gravity.

4. Proceed until the next 5° rise in temperature occurs, and remove fraction as before. Continue fractionating every 5° rise, until
   (*a*) the total distillate reaches 95 c.c. *or*
   (*b*) the temperature reaches 200° C.

5. Measure the residual liquid in the flask, and determine its refractive index (and) or specific gravity. Compare the results with those obtained in the same apparatus with a sample of genuine gum-turpentine.

   Owing to the varying factors in distillation, as height of neck, capacity of flask, temperature of room, rapidity of distillation, heat of flame, etc., it is advisable for each operator to use his own standard determinations.

The following tables show typical examples of fractionation:

A. *Genuine gum-turpentine, ref. index* $1\cdot4708$ $[20°\ C.] = 67\cdot6$ Z.B. *Commences to distil at* $156\cdot5°\ C.$

Fractions. °C. Limits of temperature	Volume of fraction. c.c.	Total distillate	Refractive index. 20° C.	Zeiss butyro-refr.
$156\cdot5°—160°$	$83\cdot5$	$83\cdot5$	$1\cdot4700$	$66\cdot3$
$160°\ —164\cdot5°$	$11\cdot5$	$95\cdot0$	$1\cdot4721$	$69\cdot6$
Residue	$5\cdot0$		$1\cdot4839$	$89\cdot8$

95 c.c. $[95\ \%]$ at $164\cdot5°$ C.

B. *"Russian turpentine." Commences to distil at* $159\cdot5°\ C.$

Fractions. °C. Limits of temperature	Volume of fraction. c.c.	Total distillate
$159\cdot5°—160°$	$4\cdot5$	$4\cdot5$
$160°\ —165°$	$29\cdot4$	$33\cdot9$
$165°\ —170°$	$36\cdot9$	$70\cdot8$
$170°\ —175°$	$8\cdot5$	$79\cdot3$
$175°\ —180°$	$8\cdot6$	$87\cdot9$
$180°\ —185°$	$6\cdot7$	$94\cdot6$
$185°\ —186\cdot3°$	$0\cdot4$	$95\cdot0$
Residue	$5\cdot0$	

95 c.c. $[95\ \%]$ at $186\cdot3°$ C.

C. *"Wood turpentine." First drop at* $153\cdot0°\ C.$

Temp. of fraction	Volume. c.c.	Total distillate. c.c.
$153°—155°$	$1\cdot1$	$1\cdot1$
$155°—160°$	$48\cdot3$	$49\cdot4$
$160°—165°$	$33\cdot4$	$82\cdot8$
$165°—170°$	$8\cdot2$	$91\cdot0$
$170°—175°$	$3\cdot8$	$94\cdot8$
$175°—176\cdot5°$	$0\cdot2$	$95\cdot0$
Residue	$5\cdot0$	

95 c.c. $[95\ \%]$ at $176\cdot5°$ C.

D. *Petroleum distillate,* S.G. $0\cdot8102.$ *Commences to distil at* $127°\ C.$

Limits of temp. of fraction. °C.	Volume of fraction. c.c.	Total vol. of distillate	Ref. index. 20° C.
$127°—130°$	$0\cdot5$	$0\cdot5$	$1\cdot4146$
$130°—135°$	$0\cdot7$	$1\cdot2$	$1\cdot4153$
$135°—140°$	$1\cdot3$	$3\cdot5$	$1\cdot4162$
$140°—145°$	$2\cdot7$	$6\cdot2$	$1\cdot4173$
$145°—150°$	$2\cdot5$	$8\cdot7$	$1\cdot4180$
$150°—155°$	$3\cdot4$	$12\cdot1$	$1\cdot4198$
$155°—160°$	$3\cdot5$	$15\cdot6$	$1\cdot4209$
$160°—165°$	$3\cdot1$	$18\cdot7$	$1\cdot4237$
$165°—170°$	$2\cdot9$	$21\cdot6$	$1\cdot4254$
$170°—175°$	$2\cdot7$	$24\cdot3$	$1\cdot4263$
$175°—180°$	$2\cdot7$	$27\cdot0$	$1\cdot4287$
$180°—185°$	$2\cdot5$	$29\cdot5$	$1\cdot4311$
$185°—190°$	$2\cdot8$	$32\cdot3$	$1\cdot4342$
$190°—195°$	$2\cdot7$	$35\cdot0$	$1\cdot4365$
$195°—200°$	$2\cdot7$	$37\cdot7$	$1\cdot4369$
Residue	$63\cdot3$		$1\cdot4557$

$37\cdot7\ \%$ distilled at $200°$ C.

E. *Turpentine A + 10 per cent. petroleum distillate D.*
S.G. 0·8626. *Commenced distilling at* 127·6° C.

Limits of temp. of fraction. °C.	Volume of fraction. c.c.	Total vol. of distillate	Ref. index. 20° C.
127·5°—130°			
130° —135°			
135° —140°	0·2	0·2	1·4345
140° —145°			
145° —150°			
150° —155°	0·3	0·5	1·4520
155° —160°	29·6	30·1	1·4668
160° —165°	49·3	79·4	1·4683
165° —170°	6·7	86·1	1·4695
170° —175°	3·6	89·7	1·4703
175° —180°	1·8	91·5	1·4706
180° —185°	1·3	92·8	1·4708
185° —190°	0·6	93·4	1·4697
190° —195°	0·9	94·3	1·4698
195° —196·5°	0·7	95·0	1·4693
Residue	5·0		1·4837

95 c.c. [95 %] at 196·5° C.

In the instance given, the amount of adulteration can almost be arrived at by inspection of the table. It is however safer to calculate the percentage present in each fraction by the ref. index, then the actual volume of adulterant per fraction, the total of these giving the total of petroleum present.

Solubility in Fuming Nitric Acid

This method, due to *Burton*, and modified by *Rothe, Marcusson*, and *Winterfield*, gives a reliable estimate of the mineral hydrocarbons present as adulteration. It depends upon the formation of soluble resinous compounds by the action of fuming nitric acid upon turpentine, which do not separate from the acid upon dilution. Mineral naphthas are either unacted upon by the fuming acid, or, if aromatic or olefin hydrocarbons are present, nitro-bodies are formed which can be dissolved out by ether on dilution of the fuming acid with water.

Burton's Method (modified)

Object. The estimation of petroleum or benzene distillates as adulterants in turpentine.

Apparatus and Reagents required

[A] 100 c.c. flask, with neck graduated for 10 c.c. in $\frac{1}{10}$ c.c. divisions.
[B] 500 c.c. flask, with graduated neck as above.
[C] Separating funnel 10 c.c. capacity with long delivery tube, narrowed at end.
[D] Separating funnel about 150 c.c. capacity.
[E] Fuming nitric acid (S.G. 1·52).
[F] Nitric acid conc.
[G] Ether.
[H] Alkali soln. (10 grms KOH in 100 c.c. water + 10 c.c. alcohol).
[I] Ice and salt freezing mixture.

Procedure

1. Place in 100 c.c. flask [A] 30 c.c. of fuming nitric acid [E] and cool flask in freezing mixture [I] to − 10° C.

2. In separating funnel [C] place 10 c.c. of the sample of turpentine, and run into the fuming acid drop by drop with continuous agitation, taking at least 45 minutes for the complete addition.

3. Stand in the freezing mixture 15 minutes, and then add sufficient conc. nitric acid [F], previously cooled to − 10° C., to bring the unattacked oil up into the graduated neck. Allow to stand in the freezing mixture until the oil in the neck is at the room temperature. Read off volume ($= a$).

4. Transfer the whole liquid to separating funnel [D], and after settlement run off lower layer (fuming nitric liquid) into flask [B] containing 150 c.c. water. (The temperature rises, and separation of oil occurs if naphthas are present.)

5. Warm for 15 minutes to complete the reaction of the nitric acid and the turpentine: cool; add 100 c.c. ether and shake, settle, run off aqueous layer and wash ether layer with water till free from acid, then with weak alkali [H] and again with water.

6. Dry the ethereal solution with a little granulated $CaCl_2$, evaporate off, and weigh residue. This consists of nitro-derivatives, and as the S.G. is approximately 1·15, it is calculated to c.c. ($= b$).

7. The total volume of mineral hydrocarbons present is ($a + b$). The a fraction should be examined for ref. index, S.G. and B.P.

Petroleum spirit. Source		S.G.	Soluble fuming nitric	Insoluble fuming nitric	S.G. insoluble portion
American	...	0·734	10	90	0·72—0·73
Russian	...	·799	10	90	0·78
		·790	8	92	
E. Indian	..	·782	22	78	0·76—0·77
		·803	40	60	
Galician	...	·760	17	83	0·74—0·75

Remarks

Marcusson (*Chem. Zeit.* 1912, **36**, 413) suggests to add conc. sulphuric acid to the separated nitric acid solution to cause nitro-derivatives to rise, so that the volume may be directly read off. Also, in case over 50 per cent. of benzene derivatives are present, to omit the 15 minutes heating (5) and divide final weight by factor 1·15, since results will then be too high.

Example

Sample of turpentine containing 10 per cent. of E. Indian petroleum spirit.

Vol. of liquid insol. in fuming nitric = 0·4 c.c. (= 4 per cent.).

Wt. of ether extract = 0·79 grm

$$= \frac{0·79}{1·15} \text{ or } 0·70 \text{ c.c.} (= 7 \text{ per cent.}).$$

Total of spirit found = 11 per cent. (error + 1·0 per cent.).

B. PINE OIL

Freshly distilled samples should be employed.

Wolff's Test [1]

Boil 5 c.c. of the sample with 5 c.c. of nitrobenzene, add 2 c.c. of 25 per cent. HCl and cautiously heat to boiling for 10 seconds (pine oils react vigorously). In the presence of pine oil the oily layer is coloured brown and the acid brown to black. In its absence the oil is a pale yellowish green and the acid a darker green.

Piest's Test [2]

Shake 5 c.c. of the turpentine with 5 c.c. of acetic anhydride; add 10 drops conc. HCl; cool; add further 10 drops HCl. On allowing to stand at room temperature, darkening occurs in presence of pine oil, and in its absence the turpentine remains colourless.

C. ROSIN SPIRIT

Grimaldi's Test [3]

In the fractional distillation (page 214) the distillate is collected in fractions of 3 c.c., or at intervals of 5°, whichever yields the smaller volumes, and is shaken in test-tubes with conc. hydrochloric acid and granulated tin, and the tubes kept at 100° C. In the presence of rosin spirit (pinolin) an emerald green coloration develops.

Zune's Test

Zune suggests to distil 100 c.c. of the sample, collecting first 25 c.c. of distillate separately, distilling a further 5 c.c., and examining the first fraction and residue for ref. index. With pure turpentine difference between the two is less than ·004. In the case of rosin oil, even one per cent., it is over this figure.

D. CHLORINATED HYDROCARBONS

These have been used to mask additions of light petroleum distillates, and for their supposed property of rendering the turpentine non-inflammable.

Their presence may be readily detected by the green coloration of a flame produced by a piece of copper wire, dipped in the liquid, and held over a bunsen flame.

The amount present may be estimated by the method of Carius for chlorine ($CCl_4 = 92·2 \%$ Cl).

SUMMARY OF TESTS FOR ADULTERANTS

Russian Turpentine
> Odour.
> Fractionation test.

Wood Turpentine
> Fractionation test.
> Lower iodine value.

[1] *Farbenzeitung*, 1912, **17**, 21.
[2] *Chem. Zeit.* 1912, **36**, 198.
[3] *Ibid.* 1907, **31**, 1145.

Pine Oil

>Lower iodine value.
>*Liebermann-Storch* test (page 222).
>*Wolff's* test.
>*Piest's* test.

Rosin Spirit

>Lower iodine value.
>Fractionation test (lower B.P. fractions).
>*Grimaldi* test.
>*Zune's* distillation test.

Petroleum Distillates

>S.G.
>Ref. index lower.
>Lower iodine values.
>Solubility tests.
>Fractionation test.
>Solubility in fuming nitric acid test.

Benzoles

>As for petroleum distillates.

Chlorinated Hydrocarbons

>S.G.
>Fractionation test.
>Copper → green flame.
>Estimation by *Carius'* method.

SECTION VIII

INTERPRETATION OF RESULTS

SECTION VIII

INTERPRETATION OF RESULTS

§ 1. The rational method of examination of oils, fats, and waxes—as previously pointed out (Vol. I, page 64)—would be to isolate the constituent glycerides, esters, etc., and determine the identity and the proportions present of each, or failing this, to split up the various glycerides or esters, and examine the individual fatty acids and alcohols. Only in a few cases is it practicable to attempt a separation of the component parts as a method of analysis, e.g. in the following:

> Insoluble bromide value—"Hexa" and "octo" unsaturated acids separated as bromides.
>
> Reichert-Meissl-Polenske-Kirschner—Partial separation of volatile acids.
>
> Unsaponifiable matter—Separation of alcohols and hydrocarbons.
>
> Renard ⎫
> Bellier ⎬ tests—Separation of fatty acids of high molecular weight.
> Tortelli ⎭

Reliance must therefore be placed mainly upon the comparison of the physical and chemical data of the oil, fat, or wax under examination with those yielded by pure specimens of the same.

§ 2. It is obvious, therefore, that a consideration of the first importance is the extent of variation shown by samples of the same oil, fat, etc. from different sources and under differing conditions. It will be advisable to consider this question in some detail.

§ 3. Effect of Differing Varieties of Plant or Animal

Oils derived from differing varieties or species of the same plant or animal exhibit undoubtedly slight differences in chemical and physical properties, although our knowledge of this matter is at present very imperfect. A recent notable discovery in this connection is the enormous differences in the chemical composition of oils from various species of sharks[1].

---

[1] The *squalidae* are found to yield oils very rich in an unsaturated hydrocarbon named "Spinacene." Chapman, *Analyst*, 1917, **42**, 161.

§ 4. Effect of Climate

The effect of the climate on the oil etc. yielded by an animal or plant is probably slight as far as the chemical and physical data are concerned. The *flavour* of an oil, however, is undoubtedly influenced by the climatic conditions obtaining. Thus olive oils from various countries differ greatly in their palatability, although differences in the variety (§ 3) may probably also have some bearing on this question. The factors operating in this case are the differences of soil, mean temperature, and rainfall. As an example from animal sources, the difference in chemical composition of butterfat from Irish butter may be cited.

§ 5. Effect of Food

Food, in the case of animal oils and fats, is a very important modifying factor. In the case of animals NATURALLY FED there are variations in the character, variety, and amount of the food taken, and the percentage of moisture taken with it. The different results given by butterfat obtained from various parts of the world are probably due chiefly to the character of food taken. The food of FISH-EATING MAMMALS undoubtedly influences the content of unsaturated glycerides in the oil derived therefrom. Another instance is the great variation in the composition of BEESWAX from different parts of the world, due principally, doubtless, to the varying character of the flora of the countries concerned.

When animals are fed upon an ARTIFICIAL DIET, not only is there often a marked difference in the flavour of the oil or fat from such animals, but the character of the glycerides is frequently modified. Thus, in addition to the alteration in flavour of the butter derived from cows put upon an artificial diet, there is an alteration in the percentage of volatile fatty acids yielded in the Reichert-Meissl test. Where the food of mammals contains glycerides differing in character from those yielded by the animal itself, a proportion of these glycerides occur in the body and milk fats of such mammals[1].

The absorption of other substances accompanying fats, such as *sesamin* from sesamé oil, and the substance in cottonseed oil giving the *Halphen reaction* resulting in the fat from the animal so fed giving the same characteristic colour reactions, is now well established.

Fortunately for the analyst, the STEROLS present in the oils used as fodder do not appear to become absorbed in this manner: thus none of the phytosterols have been detected in the tallow from cattle fed on vegetable products, or in lard from hogs fed with cotton cake. This test[2] is therefore still available for determining the animal or vegetable source of an oil or fat.

---

[1] According to recent researches, the oils taken as food by mammals are partially hydrolysed in the body and re-esterified as *mixed glycerides*, having fatty acid groups proper to the normal fat of such mammals.

[2] Phytosteryl acetate test: page 178.

§ 6. In addition to the variation in natural conditions, which may modify the character of the oils, etc., there are variations due to the manufacturing processes employed, from the treatment of the animal or plant tissue to the final sale of the oil to the general public. In the course of the treatment which the oil undergoes, it may also suffer inadvertent contamination, and it is often important to distinguish this from wilful sophistication.

§ 7. Purity of the Source

Although doubtless in many cases oil is knowingly adulterated by mixing cheaper fatty material with the original product, there are also some instances in which such admixture is due to ignorance, inadvertence, or to unavoidable natural conditions. Thus, linseed oil is, from some sources, regularly contaminated with oil from other seeds, and fish livers are seldom obtained from a single variety of fish.

"Japan fish oil" may be from the Japanese herring, or the Japanese sardine, or from both.

§ 8. Freshness of the Oil-yielding Material

To this factor is due the greatest part of the variation in colour, odour, and flavour of oils and fats. The chemical character is also liable to great alterations. Thus the free fatty acidity of an oil or fat is largely a measure of the freshness of the fruit, seed, or, in the case of animals, of the fatty tissue from which it has been derived. In many cases, oxidation of the glycerides is produced, resulting from the same cause, and, if decomposition of the original tissue has occurred to a large extent, the glycerides may have been more or less broken down, resulting in the formation of lower fatty acids, aldehydes, etc. Ptomaines, and other organic bases, occur in oils from putrid fish and blubber of marine animals.

§ 9. Method of obtaining Oil, Fat, etc.

In the case of oils and fats obtained by expression, the colour and flavour of the oil is greatly influenced by the TEMPERATURE of the expression, length of time taken for "cooking" the seed, etc.; the cold expressed oils being always lighter, sweeter and purer.

Albuminous impurities accompany most expressed oils to a greater or less extent, though these are generally to a considerable degree removed by after refinement.

Extracted oils may contain some of the solvent used for extraction, especially if a complex solvent, such as petroleum ether, has been used for this purpose. "Sulphur olive oils" contain sulphur from the carbon disulphide employed in the extraction of the olive "marc."

Oils obtained by boiling animal tissue with water, especially when in a putrid condition, as in the case of lower grades of fish oils, are liable to contain excess of moisture and much albuminous impurities. Vegetable waxes, which are mostly obtained in this manner, are seldom free from moisture, and Carnaüba wax usually contains, in addition, much admixed foreign matter, such as fragments of vegetable refuse, which has risen to the surface of the water with the wax.

§ 10. Method of Refinement

Refinement which aims at the removal of impurities, such as acid or alkali treatment, has no effect whatever on the oil, fat or wax, except in the removal of free acidity in the latter case, and in the former, the increase in acidity due to slight hydrolysis which may occur in some cases with the acid treatment.

Bleaching and deodorisation processes, on the other hand, often affect the glycerides to a slight extent. Thus, when air or ozone is employed, there may be some oxidation of the unsaturated glycerides, and this may also result, to a small degree, from the treatment of oils with dichromate and permanganate. Chlorine bleaches may slightly chlorinate the oil as well as oxidise it. Fuller's earth treatment has no effect whatever upon the glycerides. Oils which have been chemically bleached tend to darken on keeping or on saponification, since the colouring matters are not necessarily removed or destroyed, but only oxidised or reduced, and may be reformed under suitable conditions. The same remark applies to most "deodorised" oils.

Distillation—often resorted to in order to refine low grade oils and fats, or for recovery of these from the sludges and "foots" resulting from the refinement of oils—has the effect of more or less completely splitting the glycerides, with the production of fatty acids. In addition to this, the high temperatures employed always result in a certain amount of decomposition of the fatty acids with the formation of hydrocarbons. Such materials would therefore give high yields of "unsaponifiable matter."

Hydrogenation (*Hardening*)

When an oil is treated with hydrogen under pressure, in the presence of a catalyst, not only are the unsaturated glycerides more or less reduced to saturated fats—resulting in a hard substance—but the colouring matters are also bleached and the odour almost completely removed. This treatment therefore produces the greatest modification in oils of any technical process, although it is hardly correct to use the term "refinement" for hardening an oil, since the oil ceases to exist as a liquid glyceride. In hydrogenation the analytical data suffer modification in the direction indicated under :

1.	Sp. Gr.	little or no change.
2.	M.P.	raised.
3.	S.P. fatty acids	raised.
4.	Ref. Index	lowered.
5.	Viscosity	rendered inapplicable.
6.	Solubility (true Valenta)	raised (solubility reduced).
7.	Polarimeter	?
8.	Iodine value	decreased.

9.	Elaïdin	rendered inapplicable.
10.	Saponification value	practically unaltered.
11.	Insoluble Br. value	reduced to zero.
12.	Reichert-Polenske value	practically unaltered.
13.	Acetyl value	usually reduced (OH split off).
14.	Acid value	unaltered.
15.	Unsaponifiable	unaltered.
	Phytosteryl acetate test	unaltered[1].
	Colour tests	rendered negative in most cases.
	Renard test	modified, though still applicable.

Little guide as to the original source of a hydrogenated oil is available, and it may be difficult in many cases to distinguish a hardened from a natural fat. The detection of traces of Nickel (page 157) will generally prove its presence, but improved methods of work have made it possible to remove all traces of the Nickel catalyst, unless the oils have a high acidity.

Hardened oils apparently saponify with alkalis more readily than the natural fats, and the authors hope to develop a method for their recognition based upon this observation. There are also certain somewhat fine distinctions in the external physical characteristics of hardened fats which are a help to recognition to the initiated, but these are difficult to describe in set terms. A further point is that the crystals obtained by the evaporation of an ethereal solution often differ from the stearin obtained from beef fat.

The ODOUR of the original oil, although usually unrecognisable in the fats themselves, may sometimes be detected by the odour developed on saponifying and liberating the fatty acids with dilute sulphuric acid (page 272 (i)).

It may also be pointed out that the oils most commonly hydrogenated, i.e. the drying, semi-drying and marine oils, all contain esters of unsaturated fatty acids having 18 carbon atoms; these, on hydrogenation, give stearic acid in much greater proportions than are present in the majority of vegetable fats. It follows, therefore, that if, after proving that a given sample is a *vegetable* fat by means of the phytosteryl acetate test (page 178), a large percentage of stearic acid is found (page 173) it may be taken as probable that a hardened fat is under consideration. ·

In the data above it is seen that the saponification value is practically unaffected in hardening, but unfortunately, in most cases, this gives little guide, as the oils usually chosen for hardening contain glycerides which are normal in molecular weight. The same remarks apply in the case of the Reichert-Meissl value. In the case of hardened RAPE OIL this would be identified by the sapon. value and Valenta test, as well as by the Renard test (the erucic acid being converted into behenic acid). CASTOR OIL, on hydrogenation, loses the hydroxyl radicle of the ricinoleic group, and to the extent to which this occurs, becomes indistinguishable

---

[1] Prolonged treatment at high temperatures appears to reduce the amounts of sterols present, especially in the case of cholesterol (Marcusson, *Zeit. angew. Chem.* 1914, **27**, 201).

from other hardened oils. In practice, however, the acetyl value would never be entirely negative, and this test would therefore still distinguish the fat.

There can now be little question that hardened oils can be legitimately employed to replace natural fats for all practical purposes, including the manufacture of edible products. It is usually necessary, in the last case, to employ a judicious mixture, since the hardened fat is normally too hard for use alone, and further, it has not quite the same "*texture*" as a natural fat. In soapmaking it has been found that much larger proportions of rosin may be satisfactorily employed when using the "hardened" stearines to replace tallow or palm oil.

The question of hydrogenation does not affect the analysis of oils *as oils*, since PARTIAL HYDROGENATION is not in practice carried out[1]. It may therefore be fairly safely assumed that oils have not been refined in this manner.

§ 11. Effect of Age

The keeping of oils and fats for a long period without any chemical or physical change resulting is possible providing that

(i) all traces of natural fat-splitting ferments, normally present in oils, and derived from the seeds, are removed or destroyed ;

(ii) the oils are kept free from all traces of moisture, and air and light are rigorously excluded.

Otherwise free fatty acids may be produced, and a certain amount of decomposition occur (as well as oxidation) resulting together in the production of *rancidity* (see Vol. I, page 16). Some oils are more liable to these changes than are others. Bleaching and deodorisation (apart from the development of odour due to rancidity) always occur to some extent, especially if light is freely admitted.

§ 12. Special Treatment

Oils are modified in several important respects by various technical processes for special purposes. A detailed description of the methods employed does not concern us here, but a brief indication of the changes effected in the more important cases may be of service :

(*a*) LITHOGRAPHIC VARNISHES (Polymerised oils).

These are produced by heating the oils to high temperatures (250°—300° C.). In the case of linseed oil, which is chiefly used, the s.g. is increased to 0·96—0·98, and the iodine value decreased to 120—80, according to the temperature and the length of treatment. The oils are transparent, viscid liquids, and the exact nature of the changes which occur is not understood, but polymerisation of the glycerides proceeds to a considerable extent.

---

[1] Deodorisation, one of the chief objects in oil refining, is apparently not effected in hydrogenation until a high degree of *hardening* has taken place.

(*b*) BOILED OILS.

Boiled oils are produced by heating suitable oils to about 150° C. with small quantities of substances termed "driers," whereby the oil acquires the property of rapidly drying to an elastic skin.   Nearly all commercial boiled oils are prepared from linseed oil, although other drying oils may be used.   The driers added may be the following, or similar compounds : lead and manganese oxides, acetates, oxalates, borates, "rosinates," and insoluble fatty acid salts.   Calcium salts are also used to some extent, and various mixtures of all the above.   Many substitutes for genuine boiled oil are met with, usually mixtures of rosin and fish oils, etc.

As in the case of polymerised oils, the S.G. is increased (to ·947—·980) and the process renders the oil thick and viscous.   The iodine value is reduced to as low as 70 in some cases.   The acid value also increases. Polymerisation and some decomposition occur, but the exact changes undergone by the oil require further elucidation.   Boiled oils may be distinguished by the ash they yield on incineration, due to the presence of the drier.   The practical test concerns the length of time taken for a thin layer to dry to an elastic skin, and the quality of the skin produced.

(*c*) BLOWN OILS.

The blown oils are produced by blowing air through the heated oil. They are used as lubricants usually in admixture with mineral oils, owing to their high viscosities.   They simulate castor oil, and have high acetyl values, by means of which they may be recognised.

(*d*) SULPHONATED OILS (Turkey-red oils).

These are produced by the action of concentrated sulphuric acid on castor, olive, cottonseed and other oils.   They are used extensively in the process of dyeing, and for producing oil emulsions with water.   The percentage of sulphonated fatty acids may be obtained[1] by boiling with dilute hydrochloric acid (1—5) for 40 minutes with constant agitation. The fatty layer is dissolved in ether, and the sulphuric acid in the aqueous liquid pptd. with barium chloride.

§ 13.   In addition to the natural variations in oils, fats, etc. and those produced in process of manufacture, there may be **inadvertent contamination or modification** of oils.   This may occur in many ways. Thus, at any time during the handling of the oil, it may be placed in tanks containing other oils, or into barrels and casks which have been inadequately cleansed ; or the oil may have been settled, and the liquid portion run off the top.   The latter is quite a common occurrence, and accounts for the very variable stearine content of technical samples of oils.

§ 14.   Notwithstanding the foregoing causes, which operate to produce variation in oils and fats, there is a general agreement amongst the

---

[1] Herbig, *Chem. Revue*, 1902, 5.

majority of samples of an oil or fat, and it is possible to consider a certain definite figure as of greater probability in a chemical or physical test than any other. The authors have attempted to indicate this figure, terming it the *average* value, and this should be the value of a "typical" specimen of the oil, etc.

§ 15. It is however of importance to know what are the outside limits of variations which have been recorded in the figures for oils, etc., as it is *à priori* improbable that a given sample should be found to vary outside the limits already ascertained by other observers. The authors have therefore given, in a graphic manner[1], the average values of the more important analytical data and the extreme limit of variations recorded, rejecting those which, in their judgment, were unreliable.

The weeding out of untrustworthy data is obviously a matter of time and of long experience, and it is possible that the authors have admitted too wide variations in the recorded figures, but it is, in their view, wiser in such a case to err on the side of conservatism.

Under the descriptions of the various oils and fats, the "usual variations" are also given where these are known.

§ 16. The method of presentation in the diagrams is as follows: The various values are represented in the form of a vertical scale, ranging from the lowest to the highest known figures for oils, fats, and waxes. The different classes of oils, etc. are distinguished by different colours, and the AVERAGE value for each oil, etc. is indicated by a horizontal coloured line, drawn opposite its position in the vertical scale. The limit of variation is shown by a thickened vertical line in colour. Thus, to take an instance at random, the titer test of rape oil (Vol. I, page 70) has variations recorded from 11° to 14°, but the value of a typical sample is 12·5°, as indicated by the horizontal line. (The figures on the scale can be readily ascertained by placing a straight edge horizontally, as indicated by the dotted lines.)

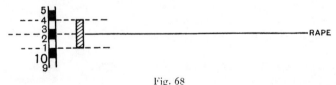

Fig. 68

§ 17. The simplest case with which the analyst has to deal, is to discover the **identity** of a sample of oil, fat or wax, of unknown origin, assuming that the sample is unadulterated. This case is fully dealt with in the following section of this book (page 267), where a tabular scheme is given for guidance (especially applicable to the student).

[1] Vol. I, Chapters IV and V.

§ 18. A more usual problem is to ascertain the purity, or the extent and character of adulteration if present, of a sample of a known oil, etc. reputed to be genuine. The following suggestions may be of service :

A. Ascertain if the sample agrees in general external character with the description given in standard text-books (or in Vol. I, Section IV). Rub a little between the hands and test the ODOUR : if an edible oil, try the FLAVOUR. Note the CONSISTENCE and, if an oil, the amount of "stearine" present.

B. Consult the trade papers and find the ruling market price of the substance, and make a note of what possible adulterants are equal to and less than this in price. (This is usually a most valuable guide.)

C. The detailed description of the oil, etc. will generally show the best tests to employ : but the following may be given as of greatest general utility as SORTING TESTS :

    1. Vegetable and animal oils—iodine value or refractometer.

    2. Vegetable and animal fats—refractometer or saponification value.

    3. Vegetable and animal waxes—melting point.

    4. Hydrocarbon and mineral oils—specific gravity.

D. Having ascertained this preliminary figure, turn up the coloured diagram (Vol. I, Chapters IV and V) concerned, or the figure given under the detailed description. If the figure obtained *coincides with the average value* there given, there is an initial presumption of purity. It is well not to rely on this first result (since mixtures may be readily made to comply with one test), and we suggest to proceed as follows :

Ascertain, by means of the special diagrams, which oils, coming into practical range as adulterants (B above), are adjacent to the oil in question. Make a note of each of these and turn up the other diagrams, noting the one in which the figures for these oils show the greatest average divergence from the sample in question. Choose this test, and compare the results as before. If the figure obtained is again identical with or close to the average figure, there is a greatly increased presumption of purity. In particular cases, a third figure should be ascertained, and negative values also obtained in special tests for specific adulterants.

Example 1. A sample of olive oil, reputed pure

The iodine value was found to be 83·5.

This coincided with the figure for a typical oil. There was therefore a pre-liminary indication of purity.

The likely adulterants, in consideration of current market values, were arachis, cotton, sesamé, rape and the drying oils.

The drying oils were definitely excluded, since they would raise the iodine value to a great extent, unless the addition were masked by some oil with lower iodine value.

Reference was made to the other tables, and the relations of these oils to olive was noted on each.

The *titer test*, although subject to great variation, showed the widest differences, if *sesamé oil* were neglected. It was therefore chosen.

The titer test gave $21 \cdot 5^\circ$ C. which was very close to the typical value for olive oil.

Sesamé oil was definitely excluded by the Baudouin test (page 135). In particular cases the s.g. or ref. index might also have been ascertained, and the following tests made :

> Halphen for cottonseed (page 133).
> Bellier for arachis (page 140).
> Valenta (true) for rape (page 79).

The oil was therefore reported as pure.

§ 19. If the second test made gives a figure which is considerably above or below the average value, although it may not fall outside the limits of recorded variations, proceed as follows :

Mark the position of the figure obtained for the oil, etc. on the coloured diagram, and make a note of the probable adulterants influencing the figure. These will be either below or above the oil according as the figure obtained is less or more than the average value. The possibilities are :

(1) The oil, etc. is exceptional in character.

(2) An adulterant is present.

(3) A mixture of adulterants is present.

Proceed now as in the previous example, consulting each diagram in order to see which value will give the greatest difference in figures between the genuine oil and the suspected adulterant. If the result of this test yields also a figure diverging widely from the average value, adulteration, although not proved, becomes more likely. In such a case, it is well to apply distinctive tests for the suspected adulterants. When the presence of a single other oil, etc. is proved, the amount present may be approximately calculated by simple proportion, taking the average value of the pure oil and of the adulterant as the basis for calculation.

This is not possible in the case of temperature figures, e.g. M.P., titer, etc., although some idea may be obtained even in such cases by the construction of tables showing the melting points of mixtures of glycerides, etc.

It must be admitted, of course, that the above gives only the *most probable* extent of the adulteration, and in cases where the indicated adulteration is small it will be well to make due allowance in order to be " on the safe side."

Example 2. A sample of lard, reputed pure

The refractometer figure was $50 \cdot 2$ (scale reading) at 40° C.

This was practically the reading of a typical oil.

The diagram gave the following information :

> If adulterated, the adulterant must be of similar refractive index to pure lard, otherwise this figure would be affected.

> Assuming a possible variation of $1 \cdot 5$ degrees scale reading on either side for pure lard, and a percentage of adulteration of not less than 5 per cent., it will be seen that oils, etc. giving refractive figures above 60 and below 40 need not be considered.

This portion of the diagram is here reproduced. Those oils which give results closest to lard are the most likely adulterants. Taking those of lower figures, we get in order of probability :

1.   Illipé.
2.   Japan wax.
3.   Mowrah.
4.   Tallow.
5.   Cacao.
6.   Sperm oil.
7.   Borneo tallow.
8.   Porpoise.
9.   Chinese veg. tallow.
10.   Butterfat.
11.   Beeswax.
12.   Palm oil.

Of these, we will assume that 5, 8, 10, and 11 are excluded as being higher in price, so that no advantage would be gained by adulteration[1].

1, 2, 3, 6, 7, 9, 12 are all less likely than 4, because as usually refined they would not make very satisfactory adulterants. 6 may be excluded altogether for this reason; the remainder, 1, 2, 3, 4, 7, 9, 12 are therefore all possibilities.

Consider now those oils, etc. which give figures higher than lard. Referring to the diagram these are, in order of probability :

13.   Dolphin.
14.   Neat's foot.
15.   Olive.
16.   Arachis.
17.   Shea.
18.   Almond.
19.   Cottonseed.
20.   Sesamé.
21.   Rape.
22.   Maize.

Of these, 13, 14, 15, and 18 were found to be quoted much higher in the current price lists. They were therefore excluded from likely consideration.

Turning now to the other diagrams, it was seen that the *titer test* separates lard widely from all the liquid oils, with the exception of cottonseed oil and arachis.

The titer test was therefore made, and it gave a result of 40·5°. This indicated the absence of liquid oils, with the possible exception of cottonseed and arachis. It also excluded fats with high titer, as Japan wax. This left for consideration 1, 3, 4, 7, 9, 12, 16, 17, 19.

In general practice the likelihood of adulteration with 1, 3, 7, 9, 12, 17 may be taken as very small, although it is well always to bear the *possibility* in mind (hence the utility of the scheme, using the diagrams).

In most cases, therefore, considering the figures recorded above, the analyst would be justified in examining for cotton oil and arachis by the Halphen and Bellier (or Renard) tests, and for tallow by examination of the lard crystals, and if all gave negative results, pronouncing as *pure*.

[1] Cases of most unlikely inadvertent adulteration have occurred, and this possibility must not be lost sight of.

(A further examination would usefully include the iodine value, and the specific gravity.)

The only adulterant at all likely from the figures cited in the above example is tallow, or *beef stearine* (page 152).

§ 20. If specific tests prove the presence of more than one adulterant an exhaustive examination must be made.

Instances occur in which this is unnecessary when the adulterants concerned give in certain tests quantitative figures which are uninfluenced by the presence of the other oils. Thus:

> **Castor oil**—except in presence of blown oils, alcohols, etc. Estimated by means of *acetyl* value.
>
> **Mineral oil**—in presence of fatty oils only by *unsaponifiable*.
>
> **Linseed oil**—in absence of other drying or marine oils by means of *insoluble bromide* value.
>
> **Arachis oil**—by *Renard's* test.
>
> **Coconut oil**—in absence of palm kernel oil by *Polenske*.
>
> **Palm kernel**—in absence of coconut oil by *Polenske*.
>
> **Butterfat**—by *Kirschner* value.
>
> **Rape oil**—in oils of normal s.v. by *saponification* value.
>
> **Fatty oils**—in hydrocarbon oils by *glycerol* determination.
>
> **Rosin**—in oils, etc. by *Twitchell* estimation.
>
> **Volatile oils**—in fixed oils and fats by *distillation*.

In these, and a few other instances, the percentage of the adulterant can be readily calculated by simple arithmetical methods[1].

In other cases the percentage of adulterant present is deduced by a careful examination of the data obtained in each test, and since the result must be **largely inferential**, the diagrams will be found of great service in arriving at a sound conclusion.

§ 21. The case of a mixture of fats and oils, etc. in which neither the number of oils present nor their identity is known, is quite a commonly occurring problem in analytical practice. The elucidation requires a **mature experience** of oil analysis and the sound judgment which such experience brings. Such a case is admittedly one of the most difficult problems in the whole range of analytical practice; all the more so when it is remembered that expert adulteration is often deliberately planned to mislead the analyst.

In specific cases even an approximate analysis may be out of the question in the present state of our knowledge, but such instances are surprisingly few in practice. Of great assistance is a knowledge of the materials likely to be employed for a given purpose, and of their current value, and these two considerations more often than anything else suggest

---

[1] Thus it would be possible, by the application of the above methods, to analyse fairly accurately a mixture of any proportion of the following eight components: castor oil, paraffin wax, linseed oil, arachis oil, coconut oil, butterfat, rosin, and benzene.

the key to problems of fat analysis which might else be wellnigh insoluble.

Since, in practice, such mixtures of oils, etc. are designed for some particular use, and their intended function is readily ascertainable, it will be preferable to consider the analysis of such groups of oils, etc. in relation to their employment. The most important may be classified as under:

A. Edible oils and fats.
B. Soapmaking oils and fats.
C. Lubricating oils and fats.
D. Candlemaking fats and waxes.
E. Illuminating oils.
F. Wax mixtures and polishes.

A.   Edible Oils and Fats

The subjoined table shows the figures for oils and fats which may possibly be used for edible purposes, excluding some of the rarer fats[1].

The best initial test is the Refractive Index, followed by the Sapon. Value, for fats, or Iodine Value for oils, and by the Valenta. Specific tests for individual oils and fats may then be made, until the range of the inquiry has narrowed down to certain members of the series. Information as to the likely components of the mixture derived from experience is here of great value.

Example 1.   A salad oil

The oil had a golden yellow colour, and a pleasant flavour. The odour was suggestive of olive oil.

The refractometer reading was 57·3. (Butyro-refractometer 40° C.)

Iodine value, 92·5.

$$\text{Valenta} \left(\frac{V \times 10}{80} \right), 9\cdot2.$$

These figures, taken together with the probable constituents, and the general character of the oil, suggested olive oil, with a mixture of sesamé and arachis, or possibly cotton. The drying oils were unlikely since the iodine value and the refractive index figure would have been influenced to a greater degree, and the Valenta lowered. If sesamé oil alone were present, the Valenta figure would probably be lower, but this would be raised by addition of a little arachis.

Specific tests were now made.

Baudouin's reaction gave a very distinct coloration. Therefore sesamé present.

Bellier's test showed arachis present, and a further examination with Renard's test showed 9·8 per cent. arachis oil. The Halphen test gave a negative result; therefore no cottonseed oil present.

The figures were considered in the light of these results. Rape oil is proved absent, since, in the presence of 10 per cent. arachis, it would raise the Valenta figure to a much greater extent.

---

[1] The list includes all the oils and fats placed in the government schedule issued by the Ministry of Food. For further details of the less familiar fats, reference to Bolton and Revis, *Fatty Foods*, is recommended.

On inspection it was seen that the figure for iodine value agreed with the following composition:

$$\begin{array}{ll} \text{Olive} & 50 \\ \text{Sesamé} & 40 \\ \text{Arachis} & 10 \\ \hline & 100 \end{array}$$

Working this out for the refractometer figure, it was found to correspond closely with this composition; and the Valenta also agreed approximately. The oil was therefore reported as *containing constituents in proportions as above.*

Example 2. A sample of margarine

The clear fat gave the following figures:

$$\begin{array}{lll} \text{Refractive Index } (40^\circ \text{ C. z.b.}) & \ldots & 43\cdot2 \\ \text{Saponification value} & \ldots \quad \ldots & 230 \\ \text{Valenta } \left(\dfrac{V \times 10}{80} \right) & \ldots \quad \ldots & 5\cdot4 \end{array}$$

These figures show the presence of coconut together with other fats. The Reichert-Polenske-Kirschner was therefore obtained, and gave

$$\begin{array}{lll} \text{Reichert-Meissl} & \ldots \quad \ldots \quad \ldots & 6\cdot5 \\ \text{Polenske} \ldots \quad \ldots \quad \ldots \quad \ldots & 8\cdot5 \\ \text{Kirschner} \ldots \quad \ldots \quad \ldots \quad \ldots & 3\cdot4 \end{array}$$

It was thus obvious that a large proportion of either coconut oil or palm kernel oil was present, and the Reichert figure showed butterfat present also. The Polenske result showed obviously about 50 per cent. of coconut, and, in this case, the Kirschner gave 10 per cent. of butterfat. There was some doubt whether palm kernel was present, though the Reichert-Meissl and the Valenta pointed to coconut only. The method of Burnett and Revis was therefore tried, and this gave a turbidity temperature figure of $53\cdot2^\circ$ C. It was therefore assumed that palm kernel oil was absent[1].

Taking the figures (coconut 50, butterfat 10) as established, the refractive index showed the probability of the addition of an oil or fat of high iodine value.

The figure for the iodine value was $27\cdot5$.

This gave a figure of about 50 iodine value for the 40 per cent. remaining components of the margarine.

$$\left(\text{Butterfat 10 per cent.} = 3\cdot3 + \text{coconut 50 per cent.} = 3\cdot3 + 4\cdot4 = 7\cdot7; \right.$$
$$\left. 27\cdot5 - 7\cdot7 = 19\cdot8; \quad \frac{19\cdot8 \times 100}{40} = 49\cdot5. \right)$$

This would be quite high for a hardened fat, which was suspected to be employed; the Halphen reaction for cottonseed oil was therefore tried and this gave a positive result, proving the presence of cottonseed oil.

Assuming 30 as a reasonable iodine value for hardened oil, this gave 25 per cent. cotton in the residual fat, or 10 per cent. in the margarine. (A titer determination and the presence of a faint trace of Nickel confirmed the probability of hardened fat being employed.)

---

[1] Further confirmation of this would be yielded by comparison of the Shrewsbury-Knapp with the Polenske figure (page 150).

Table of average values of oils more commonly used for edible purposes.

Oil	1 Sp. Gr. 15° C.	3 S.P. fatty acids ° C.	4 Refractometer (Z.B. 40° C.)	6 True Valenta* — Acetic acid	6 True Valenta* — Alcohol reagent	8 Iodine value	8a Br. thermal test
LINSEED ...	0·934	19·5	73	5·9	62·4°	175—200	30°—32°
HEMPSEED	0·927	16	—	—	—	148	—
SOYA ...	0·925	21·5	63	6·7	67·0°	130	23°
POPPYSEED	0·925	16	63·5	—	—	134	22°
NIGER ...	0·927	—	63·0	6·7	60·0°	133	—
SUNFLOWER	0·925	17·5	—	7·9	64·0°	132	21·5°
MAIZE ...	0·925	18·5	60·3	8·5	68·2°	119	—
KAPOK ...	0·924	31	51·5	—	—	97·5	19·4°
COTTON ...	0·923	35	58·5	7·5	65·2°	110	23·5°
SESAMÉ ...	0·923	23	60	8·7	68·1°	105	18·5°
RAPE ...	0·9155	12·5	59	13·8	83·3°	100	16·5°
ARACHIS ...	0·9175	28·5	55·5	11·0	74·3°	90	14·5°
OLIVE ...	0·916	20	55·5	9·25	69·1°	83	—
SHEA ...	—	54	56	—	—	60	—
MOWRAH ...	—	—	48	—	—	59	—
CACAO ...	0·970	49·5	46·7	11·7	76·0°	37	—
CH. VEG. TALLOW	0·916	52·5	44	9·3	65·9°	20	—
PALM ...	0·922	44	42	11·7	68·2°	54	—
PALM KERNEL	0·926	23	36·5	3·0	40·0°	13·5	—
COCONUT ...	0·926	23	35·5	1·5	34·0°	8·5	—
LARD ...	0·936	40	50	10·9	72·7°	62	—
TALLOW ...	0·947	44	48	11·5	72·7°	43	—
OLEO OIL ...	—	—	47·5	—	—	46·5	—
STEARINE	—	—	53	—	—	10	—
BUTTERFAT	0·938	35	43·5	4·7	46·0°	33	—
HARDENED FAT ...	—	40—65	49—57	—	—	0·5—70	—

* See pages 76—86.

Oil	10 Sapon. value	11* Insol. br value	12 Reichert-M. value	12 a Polenske	Special feature	Special test	Reference Vol. I page	Reference Vol. II page
LINSEED	190	33—38	below 1	below 1	odour	M.P. of bromides	113	111
HEMPSEED	191	—	,,	,,	green colour?	,,	120	111
SOYA	192	4	,,	,,	—		121	111
POPPYSEED	192·5	nil	,,	,,	—	Bellier	122	133
NIGER	190	—	,,	,,	flavour	,,	124	—
SUNFLOWER	193	nil	,,	,,	—	unsaponifiable	124	126
MAIZE	191	,,	,,	,,	,,	Becchi	126	134
KAPOK	193	,,	,,	,,	—	Halphen, Becchi	128	133
COTTON	193	,,	,,	,,	—	Baudouin	128	135
SESAMÉ	192	3—5	,,	,,	—	Tortelli and Fortini	131	135
RAPE	175	nil	,,	,,	viscosity	Bellier, Renard	136	140
ARACHIS	193	,,	,,	,,	—	Mazzaron	138	146
OLIVE	190·5	,,	,,	,,	flavour	unsaponifiable	140	126
SHEA	184	,,	,,	,,	—		—	—
MOWRAH	192	,,	,,	,,	,,	Björklund	151	146
CACAO	193	,,	,,	,,	—	none	153	—
CH. VEG. TALLOW	203	,,	,,	,,	odour	,,	154	150
PALM	201	,,	5·2	9·8	,,	,,	158	—
PALM KERNEL	246	,,	7·5	16·5	,,	Burnett and Revis	159	147
COCONUT	257	,,	0·6	below 1	flavour	Shrewsbury-Knapp	166	152
LARD	196	,,	0·5	,,	,,	Stock-Belfield	170	—
TALLOW	195	,,	0·66	,,	—		—	—
OLEO OIL	198	,,	below 1	·55	—		—	153
STEARINE	195	,,		below 1	flavour	crystals	172	154
BUTTERFAT	227	,,	28	2·3	—	Kirschner	—	157
HARDENED FAT...	as original oil	,,	below 1	below 1	—	detection of Nickel	—	—

* See page 112.

The margarine was therefore assumed to consist of:

Butterfat	...	...	10
Coconut oil	...	...	50
Hardened fat	...	...	30
Cottonseed oil	...	...	10
			100

B. Soap Manufacture

The examination of the oils used in the composition of soap differs from most other cases in the fact that only the mixed fatty acids are available, the oils having been split by saponification. (By the term mixed fatty acids is meant the "insoluble fatty acids + unsaponifiable.") The characteristics of these are given in section V.

The soap is decomposed by means of mineral acid, carefully noting the odour emitted, the fatty acids washed and dried, being then ready for examination.

The best INITIAL TEST is the IODINE VALUE, followed by the mean MOLECULAR WEIGHT and TITER TEST. Rosin is very frequently present, and it is best to remove it altogether from the fatty acids before proceeding with the other tests. This is accomplished by using a large quantity of the mixed fatty acids + rosin and proceeding as in the Twitchell method (page 223). After neutralising the rosin acids, the esters of the fatty acids are shaken out of the aqueous soap solution with ether, the ether evaporated off, and the esters split by boiling with alcoholic potash, evaporating off alcohol, dissolving the soap paste in water and acidifying with mineral acid. The fatty acids thus obtained are free from rosin acids, but contain any unsaponifiable matter occurring originally in the rosin.

Example 1. A sample of household soap (*in bars*)

Appearance of soap—fairly firm, light yellow colour, pleasant odour[1].
Percentage of fatty acids 57·5 per cent.
Tested for rosin by Liebermann-Storch (page 222). Rosin present.
Took large quantity (30 grms) of mixed fatty acids and proceeded as above, removing the rosin acids.
Proportion of rosin by Twitchell = 22 per cent.
The fatty acids free from rosin acids gave the following figures:

Iodine value	...	...	...	38·5
Mean molecular weight		...	...	252·5
Titer test	...	...	...	33° C.

Cotton oil was present by the Halphen reaction.

The mean molecular weight shows coconut oil present to the extent of about 30 per cent. of the *fatty* acids.
A POLENSKE determination of the mixed fatty acids[2] confirmed this figure.
Tallow, or a hardened fat was present by the iodine value. The titer test was lower than would have been expected with hardened fat, considering the amount

---

[1] Most household soaps are scented with low grade perfumes (citronella, etc.).
[2] See note, page 170.

of cottonseed oil which is indicated. Assuming tallow present, the iodine value gives about 15 per cent. of cottonseed oil.

[Coconut $\frac{9\cdot7 \times 30}{100} = 2\cdot9$; $38\cdot5 - 2\cdot9 = 35\cdot6$; $35\cdot6 \times \frac{100}{70} = 50$.

Tallow at I.V. $= 42$; cotton at I.V. $= 110$.

By inspection this indicates about 15 per cent. cottonseed oil.]

It was therefore inferred that the *composition of the fatty acids was*:

Tallow ...	...	...	55
Cottonseed oil ...		...	15
Coconut	...	...	30

Adding the rosin found, we obtain:

Rosin ...	...	...	20
Tallow ...	...	...	44
Coconut	...	...	24
Cottonseed	...	...	12
			100

The above analytical data would not, taken alone, fully justify the conclusions arrived at. These were also based upon a wide knowledge and experience of the materials likely to be employed together with their relative value at the time of manufacture. Additional help was also given by the general appearance of the soap and the odour of the fatty acids. In case of doubt the insoluble bromide value would have excluded drying oils, whilst the examination of the liquid fatty acids (lead salts soluble in ether, page 171) would have given additional information as to the amount of cottonseed oil present.

This is a good instance of the value of practical experience in commercial analyses, referred to above.

Example 2. A sample of potash soft soap

The soap was transparent, light yellow and of medium consistence.

The odour was suggestive of linseed oil.

Fatty acids per cent. $= 39\cdot5$.

The mixed fatty acids (pronounced odour of linseed) gave figures as follows:

Iodine value	...	...	...	...	... 153·0
Titer test ...	...	...	...	...	... 21·0° C.
Mean molecular weight fatty acids			...	...	283
No rosin by Liebermann-Storch test.					

These figures pointed to linseed oil and another oil as present. The insoluble bromide value was ascertained.

This gave 20·2 [1].

The proportion of linseed was thus about 50—60 per cent.

Cotton was found absent by Halphen test.

Current values for oils made it probable that only *soya bean oil* would be employed besides linseed in a soft soap. This supposition agreed with the observed lack of distinctive odour and the general appearance of the soap.

The amount indicated by all the tests was about equal quantities of linseed and soya bean oil.

The composition was therefore taken as

Linseed	...	50
Soya	...	50
		100

[1] Modified method of *Gemmell*, *v.* page 110.

Table of the average values of mixed fatty acids for Hard Soap Analysis.

Mixed fatty acids of	S.G. 90°C/15.5°C	s.f. °C	Refracto-meter†† (z.B. 40°C.)	Iodine value	Mean molecular weight	Insoluble bromide value*	Stearic acid % (Hehner and Mitchell)	Iodine value liquid fatty acids	Polenske†	Turbidity test fatty acids§	Special tests	Reference Vol. II
WHALE	·860	23	44	120	290	23	prob. absent	145	—	72°	M.P. of bromides	111
LINSEED	·861	19·5	58—59	179—210	283	33—38	traces	200—210	—	69°	,,	111
COTTON	·847	35	44	105—112	275	nil	none	150	—	76°	Halphen	133
OLIVE	·843	20	41—42	86	286	,,	,,	90—105	—	78°	—	126
SHEA	—	54	34	56	—	,,	30	90	—	—	unsaponifiable ?	—
MOWRAH	·860	38	30	64	—	,,	13—25	—	—	79°	—	—
CH. VEG. TALLOW	·837	52·5	29	24	274	,,	none	97	—	83°	—	—
PALM	—	44	—	55	273	,,	0—2	95	—	43°	Shrewsbury-Knapp	147
PALM KERNEL	—	23	19—20	14	223	,,	7	45	11·0	27·5°	,,	147
COCONUT	·835	23	10—11	9·7	211	,,	5	36	17·5	86°	Stock-Belfield	152
LARD	·838	40	37	64	278	,,	6—25	90—120	—	—	—	—
BONE FAT	—	40	—	52	280	,,	about 15	—	—	83°	—	—
TALLOW	·837	44 B. / 45·5 M.	34	46 B. / 43 M.	201	,,	33—50	92	—	—	—	—
HARDENED OIL	—	30—60	—	10—50	—	,,	80—95	—	—	—	estimation of Ni	157
ROSIN	—	—	—	110 (variable)	—	,,	none	—	—	—	Liebermann-Storch	222

†† Fryer and Weston. * See page 112. † Fryer (*Journ. Soc. Chem. Ind.* 1918, **37**, 262). § See Appendix I.

Table of the average values of mixed fatty acids for Soft Soap Analysis.

Mixed fatty acids of	S.G. 99°/15·5°	S.P. 15°C.	Refractometer†† (Z.B. 40°C.)	Iodine value	Mean molecular weight	Insoluble bromide value*	Iodine value liquid fatty acids	Acetyl value	Turbidity test fatty acids†	Special tests	Reference Vol. II
MENHADEN	—	—	55—57	163	289	51·5	—	—	—	M.P. of bromides	111
JAPAN FISH	—	15—28	46—48	120—180	—	—	—	—	—	"	"
HERRING	—	—	—	140—160	292	—	—	—	—	"	"
SALMON	—	—	54	150—170	—	—	—	—	—	"	"
COD LIVER	—	13—24	53	145—175	288	35	very high	—	—	"	"
SHARK LIVER	—	—	—	—	—	18	—	—	—	"	"
SEAL	·860	15	49	143	292	29	—	—	—	"	"
WHALE	·861	23	44	120	290	23	145	—	72°	"	"
LINSEED	—	19·5 / 16	58—59	179—210	283	33—38	200—210	—	69°		
HEMP	—	21·5	57	152	—	—	—	—	75°		
SOYA	—	22	48	135	—	4	—	—	75°		
NIGER	—	18·5	48·5	135	—	nil	—	—	76°		
MAIZE	·857	35	45	122	283	—	—	—	76°		
COTTON	·853	28·5	44	105—112	275	?	150	—	87°	Halphen, Becchi	133
ARACHIS	·847	20	42	94	282	?	—	0·9	78°	Renard	144
OLIVE	·846	3	41—42	86	286	?	90—105	1·0	soluble		
CASTOR	·843	—	58	89	306	?	—	150·0	soluble	optical activity	87
"OLEINE"	·896	15—20	40—41	80	283	?	—	—	80°		
PALM	·846	44	29	55	273	—	—	1·6	83°	Liebermann-Storch	
[ROSIN]	·837	over 100	—	110 (variable)	—	—	45	—	—		222

* See page 112.   †† Fryer and Weston.   † See Appendix I.

Example 3. A sample of "soda soft" soap

This was transparent at 20° C., slightly cloudy under 15° C. and firm in consistence. Odour of linseed oil and rosin.

The fatty acids (+ rosin) were 50·5 per cent.

Rosin was present by the Liebermann-Storch test.

The Twitchell estimation gave *Rosin 15 per cent.*

The rosin acids were not separated and the following figures apply to (rosin + mixed fatty acids):

Iodine value	145·5
Titer test indefinite (on account of rosin present) ...	25°
Insoluble bromide value	22·2[1]

The last figure gave linseed about 60 per cent. present, and this presumed the presence of some other oil with iodine value about 80—100. Castor oil was suspected for other reasons. The acetyl value of the fatty acids was then obtained. This gave 35·5, corresponding to a percentage of 25 of castor oil in the fatty acids.

The iodine value obtained corresponded with this mixture of oils and rosin, and as no other oil would be likely to be employed under the existing market conditions, the *composition was taken to be*:

Linseed oil	...	60
Castor oil	...	25
Rosin ...	...	15
		100

C. Lubricating Oils and Fats

In all cases, determine first the SAPONIFICATION VALUE (page 106).

In the case of mineral lubricants only, such as are now largely used for all classes of work, the analysis will follow the lines given under *mineral lubricating oils* (page 198), and will aim at discovering the particular fractions of the distilled products employed and their source.

If the saponification value is very high, determine the percentage of UNSAPONIFIABLE MATTER (page 126). If this is below 1·5, the lubricant probably consists entirely of animal or vegetable oils, or mixtures of these. The iodine value should then be obtained, and these figures (the S.V. and the I.V.) compared with the special diagrams (Vol. I, Chapter V) or with the subjoined table. *Castor oil* may be estimated by the acetyl value, in the absence of *blown oils*. (The latter are recognised by their high viscosity and S.G. but differ from castor in being only *partially* soluble in alcohol.)

RAPE OIL is best recognised by the *Valenta* ("true Valenta," page 79), ARACHIS by the *Bellier* and *Renard* tests (page 140). Semi-drying and drying oils are undesirable constituents of lubricating oils and may be suspected by a high iodine figure.

SPERM OIL is characterised by its low S.G. and sapon. value, and by yielding little or no glycerin on saponification.

ROSIN OIL and ROSIN are detected by the *Liebermann-Storch* test,

[1] Modified method of *Gemmell*, v. page 110.

Table of average values of oils used as lubricants.

Oil	1 S.G. at 15° C.	3 S.P. fatty acids	4 Refracto-meter 40° C.	5 Absolute $\eta \times 100$ viscosity (40° C.)	5a* Viscosity ratio number	6† True Valenta $V \times 100 / 80$ (acetic)	8 Iodine value	10 Saponification value	13 Acetyl value	Special features	Special tests	Reference, vol. and page
RAPE	0·9155	12·5°	59	40	5·0	13·8	100	175	1·5	—	Tortelli & Fortini	I, 136; II, 135
ARACHIS	0·9175	28·5°	55·5	—	—	11·0	90	193	0·9	—	Bellier & Renard	II, 138; I, 140
OLIVE	0·916	20°	55·5	34	4·8	9·25 soluble	83	190·5	1·0	—	Mazzaron	I, 140
CASTOR	0·964	3°	68·9	248	15·0	—	88	180	150·0	—	optical activity	I, 143
NEAT'S FOOT	0·917	26°	52	—	—	—	70	195	0·2	—	Stock-Belfield	I, 146
LARD	0·9155	30°	41	—	—	—	85	195	0·3	—		I, 165
TALLOW OIL	0·916	32°	—	—	—	11·7	57	195	1·8	—		I, 171
PALM	0·922	44°	42	—	—	—	54	201	1·6	colour & odour		I, 154
TALLOW	0·947	44°	48	17	3·7	11·5	43	195	0·5	—	per cent. glycerin	I, 171
SPERM	0·880	11·5°	52			10·4	86	124	5·0	a liquid wax		II, 181
BLOWN RAPE	0·962	—	—	600—800	—	—	70	215	47	—		I, 136
BLOWN MAIZE	0·980	—	—	,,	—	—	90	209	63	—		I, 126
BLOWN COTTON	0·979	—	—	—	—	—	65	225	64	—		I, 216
SPINDLE. M.	·868—·891	—	—	6—30	3—5½	11—13·5 insoluble	very small	nil	nil	—		—
LIGHT MACHINE. M.	·900—·901	—	—	30—40	5½—7½	,,	,,	,,	,,	—		II, 217
ENGINE. M.	·900—920	—	—	50—120	7½—9	,,	,,	,,	,,	—		,,
CYLINDER. M.	·900—·920	—	—	200—1000	10—25	,,	,,	,,	,,	—		,,
ROSIN	0·97—1·0	—	—	—	—	—	—	—	—	—	Liebermann-Storch	II, 222

* See page 70. † See page 76.

and rosin is estimated by the *Twitchell* method (page 223). Rosin oil is recognised by its high S.G., and by the further particulars given in section VI (page 232).

Many lubricating oils are COMPOUND, that is, they are mixtures of saponifiable oils with mineral oils. These will give definite saponification values, and the mineral oil may be removed by shaking the aqueous soap emulsion, obtained on saponification, with ether, separating, and evaporating off the ether. The residual mineral oil may be examined as above. The soap solution is then decomposed by mineral acid, and the mixed fatty acids examined (page 160).

Compound lubricating fats and greases may contain alkali soaps, and insoluble metallic soaps (calcium, magnesium, etc.).

The alkali soaps may be removed by boiling with water, and shaking out the unsaponified or unsaponifiable oils with ether (after cooling the liquid). In the case of metallic soaps (recognised by incinerating the grease, etc., and examining the ash), the best method is to boil with 1 in 4 hydrochloric acid, separate off the oily layer, wash, saponify with alcoholic potash, and shake out the oil with ether. The aqueous liquid can then be decomposed with mineral acid, and the fatty acids so obtained examined. The amount present will be best gauged by estimating the acidity of the oily layer (above) after treatment of grease with HCl, and washing free from mineral acid. A further check may be made by estimating the amount of lime, magnesia, etc., but it should be remembered that there may be an excess of this present.

Railway waggon greases usually contain palm oil, or tallow, partially saponified with soda or lime, and occasionally mineral oils in addition.

Rosin is frequently present, and may be detected by the Liebermann-Storch test, and estimated by the acidity (in mineral oil only) or the Twitchell method (in saponifiable oils).

Example 1. An "engine" oil

Fairly viscous, reddish, clear oil, with strong fluorescence ("bloom"). It gave the following figures :

S.G. (15° C.)	0·9053
Viscosity (absolute $\eta \times 100$) 40° C. ...	85·5
Viscosity ratio number	9·9
Flash point (Gray)	165° C.
Saponification value	0·25
Cold test (S.P.)	− 16° C.

No rosin or rosin oil (Liebermann-Storch), no coal-tar oils (Formolite reaction). These figures showed a *somewhat heavy mineral lubricating oil, pure, and of excellent quality,* and probably of Russian origin.

Example 2. A sample of a lubricating oil for aircraft engines

This was quite viscous in appearance, almost white, and with little odour; no fluorescence ("bloom").

It gave the following figures :

Sapon. value...	...	...	...	...	181·5
S.G. (15° C.)	...	...	...	...	0·963
Unsaponifiable matter	...	...	...	...	0·85 per cent.
Viscosity ($\eta \times 100$) 40° C.		...	...	246	

The low unsaponifiable showed a saponifiable oil, unmixed with mineral oil, and the high s.g. indicated either castor oil or a *blown oil*, though the s.v. was rather low for the latter. The acetyl value was therefore determined. This gave 149·3, showing the sample to be *genuine castor oil.*

Example 3. An oil sold for motor-car engine lubrication

Fairly viscous, reddish, transparent, with strong fluorescence ("bloom").
It gave a s.v. of 26·2.
It was therefore apparently a "compound" oil.
The mineral oil was removed by shaking the soapy emulsion produced after saponification (see above) with ether.
The aqueous liquor, on decomposition with mineral acid, yielded fatty acids, giving figures (see section V) as under :

$$\text{Mean mol. wt. } 317.$$
$$\text{S.G.} \qquad 0.8432 \left(\frac{99°}{15\ 5°} \right).$$
$$\text{Iodine value} \quad 101.2.$$

This was evidently *rape oil*, and by the s.v. there was present about 15 *per cent.*
An examination of the mineral constituent, obtained by evaporation of the ether extract, showed a *typical mineral lubricating oil*, probably of American origin.

Example 4. An axle grease for waggon lubrication

Yellow colour, stiff consistence, aromatic odour.
A weighed portion was incinerated, and the ash obtained (10·63 per cent.) consisted almost entirely of sodium carbonate.
A weighed amount was then boiled with dilute sulphuric acid, the oil filtered off, dried, and weighed. This gave oil 73·5 per cent. and the acidity was 37·3 per cent. in terms of oleic.
It was then saponified and decomposed, washed and the following tests made on the fatty acids :

Titer	...	...	...	44° C.
Iodine value		..	...	49·3
Mean mol. wt.		...	...	278·5

This was apparently a mixture of tallow and palm oil, and the estimation of the stearic acid (Hehner's method, page 167) confirmed this.
The grease had therefore the *following composition* :

Tallow	36
Palm oil	37
Soda	10·6
Water	16·4
	100 0

(This is not strictly correct, since a portion (37 per cent.) of the oil existed as soap combined with the soda.)

D. Candle Manufacture

Candles are made from

    1. " Stearines " of various grades.
    2. Paraffin wax.
    3. Mixtures of 1 and 2.
    4. Beeswax.
    5. Ozokerit (ceresin).
    6. Spermaceti.

In addition, carnaüba wax, montan wax, and other substances are employed as " stiffeners."

1. " STEARINES."

This is the commercial term for solid fatty acids.   They are produced by the various processes of hydrolysis (autoclave, Twitchell, etc.) or by the "acid" process (treatment of fat with sulphuric acid), and in each case the fatty acids are usually distilled to get a white fat.   Sometimes the " saponification " stearine is sufficiently light without distillation.   The distilled fatty acids are hot-pressed to exclude the liquid fatty acids.   The acid process gives a higher yield (60—65 per cent.) of hard fatty acids, but the melting point (47—52° C.) of the " stearine " (termed " distillation stearine ") is lower.   (" Saponification stearine " has a M.P. of from 52—56° C. and a yield 45—48 per cent. is obtained.)   In the case of the acid process, hydroxystearic acid and stearolactone are amongst the products formed, also a little iso-oleic acid.   It may thus be distinguished from " saponification stearine " by the iodine values :

    Iodine value.   Distillation   stearine   15—30,
            ,,   Saponification   ,,   0·5—8·0.

Almost all fats may be employed to yield " stearines," and their value for this purpose is gauged by the commercial yield of stearine and its melting point.

Stearines may thus consist of mixtures of solid fatty acids, chiefly stearic and palmitic.   Mixtures of the acids have the following melting and solidifying points :

Stearic per cent.	Palmitic per cent.	M.P.		S.P.	
		° C.	° F.	° C.	° F.
100	0	69·2	156·5	—	—
90	10	67·2	153	62·5	144
80	20	65·3	149·5	60·3	140
70	30	62·9	145	59·3	138·5
60	40	60·3	140·5	56·5	133
50	50	56·6	133·5	55	131
40	60	56·3	133	54·5	129·5
30	70	55·1	131	54	129
20	80	57·5	135	53·8	128·5
10	90	60·1	140	54·5	129·5
0	100	62·0	143·5	—	—

Table of average values of waxes, etc.

For examination of technical wax products (candles, polishes, etc.).

Wax, etc.	1 S.G. 15·5° C.	2 M.P. °C.	4 Refractive index (z. b. 40°)	6†† Solubility alcohol reagent	8 Iodine value	10 Saponification value	14 Acid value	15 Unsaponifiable	Fatty acids %*	Ratio number	Special features	Reference, vol. and page
CARNAÜBA	·998	84	67	82°	13	80	2·5	55	48	31	characteristic odour on warming	1, 182
CANDELILLA	·972	66	62	63°	35	63	11	74	29	4·7	distinctive odour	1, 184
CANE SUGAR	·980	58	—	—	88	80	12	69	33·3	5·7	nearly black colour	—
BEESWAX	·965	63	44	76°	9	95	20	55·5	46·8	3·6—3·8	distinctive odour	1, 185
INSECT WAX	·970	81	—	insoluble	1·5	91	3	49·5	51·5	29·3	fibrous crystalline structure	1, 188
SPERMACETI	·960	44	—	44°	4	125	1	51·5	53·5	124	glistening translucent crystalline	1, 188
MONTAN WAX (ref.)†	—	77—84	—	—	—	70—80	15—20	—	11—15	3—3·5	earthy odour, crystalline structure	1, 191
MONTAN WAX (dist.)†	—	72—77	—	70°	10—15	75—89	73—85	30—45	56—64	·03		1, 191
PARAFFIN WAX	·87—·91	35—75	23—30	insoluble	nil	nil	nil	100	nil	0	vitreous transparence	1, 207
CERESIN (Ozokerit)†	·920	68—73	35—42	insoluble	nil	nil	nil	100	nil	0	plasticity	1, 208
["STEARINES"]	—	49—56	30—33	33°	0·5—30	200	198	0·5—4·0	96—99·5	·01	may contain traces of nickel, p.	II, 233 / II, 157
["HARDENED OIL"]	—	30—60	30—33	73°	10—50	198	variable	0·5—2·0	95	25—200		
[ROSIN]	1·07—1·08	over 100	—	below 15°	55—180 av. 110	147—183 av. 175	130—180 av. 164	5—15 av. 7·5	—	0·1—0·2	Liebermann-Storch test	1, 255 / II, 222
["JAPAN" WAX]	·987	52	48	76°	6	220	6—20	1·0	90	11—35	effloresces, tallowy odour	1, 156

† Fryer (average of samples examined).   †† Fryer and Weston (p. 86).   * Lewkowitsch.

The **hydrogenation** of oils by reduction with hydrogen and a catalyst is now extensively employed for the commercial production of high grade "stearine." By this method, almost any oil may be employed for candle manufacture, and practically no loss by the expression of lower melting point fractions is incurred.

2, 3. PARAFFIN WAX.

From petroleum, shale, or lignite (see Vol. I). It is seldom employed alone, being too soft, but is stiffened with 5—20 per cent. of "stearine," or ceresin, carnaüba wax, montan wax, etc. are used for this purpose. The stearine present in an ordinary paraffin candle may readily be ascertained by means of the acid value, and by the melting point (see subjoined table).

4. BEESWAX.

"Church candles" are usually made with proportions of beeswax, seldom the pure wax. Often the percentage of beeswax is stamped on the candle. Stearine, paraffin wax, ceresin, carnaüba wax and montan wax are employed in their manufacture.

(A) Melting point of mixtures of stearines with Scotch paraffin waxes [° F.][1].

Stearine[2] per cent.	Stearine 121° F. with paraffin wax of M.P. ° F.			Stearine 123° F. with paraffin wax. M.P.	Stearine 129·75° F. with paraffin wax. M.P.	
	102°	125°	130°	120°	120°	132·5°
10	100	123	128	118	118·5	130·5
20	98·5	121	125·5	116·5	116·7	128·5
30	100	119	123	114	114·5	126·5
40	104·5	117·5	121	112	112·2	124·2
50	110·5	114	118·5	110	113	121
60	111·0	111	114	109	118·7	117·7
70	113·5	107	109	113	112	119·5
80	117·5	114	115·5	118·5	124·5	125·2
90	119·0	117	118	119·5	127	127·5

The complete analysis of such composite candles is a matter of some difficulty. If only beeswax and a mineral wax or waxes are present, the percentage of beeswax may be obtained by the saponification value. The ratio number is also an important figure (see Vol. I, page 186).

5. OZOKERIT (ceresin).

This is not often employed by itself as a candle material, owing to the fact that it cannot be moulded (like pure beeswax) but must be "drawn." See Vol. I, pages 208, 245; Vol. II, page 219.

---

[1] J. I. Redwood.
[2] The paraffin present in each is the percentage of stearine less 100.

(B) *Melting points of candle material from Thuringian paraffin wax and " stearine " [° C.]*[1].

Paraffin wax. Per cent.	Of melting point. °C.	"Stearine" of melting point 54° C. Per cent.	Melting point of mixture. °C.
90·0	36·5	10·0	36·5
66·6	,,	33·3	39·0
33·3	,,	66·6	45·75
10·0	,,	90·0	51·75
90·0	37·5	10·0	36·5
66·6	,,	33·3	35·5
33·3	,,	66·6	47·0
10·0	,,	90·0	52·0
90·0	40·75	10·0	39·75
66·6	,,	33·3	40·50
33·3	,,	66·6	47·50
10·0	,,	90·0	52·0
90·0	45·0	10·0	44·0
66·6	,,	33·3	40·75
33·3	,,	66·6	48·0
10·0	,,	90·0	52·5
90·0	48·5	10·0	47·5
66·6	,,	33·3	45·0
33·3	,,	66·6	47·75
10·0	,,	90·0	52·50
90·0	50·0	10·0	49·0
66·6	,,	33·3	47·0
33·3	,,	66·6	47·5
10·0	,,	90·0	52·5
90·0	54·0	10·0	53·0
66·6	,,	33·3	49·0
33·3	,,	66·6	47·0
10·0	,,	90·0	52·5
90·0	56·5	10·0	55·5
66·6	,,	33·3	52·0
33·3	,,	66·6	47·5
10·0	,,	90·0	52·5

6. SPERMACETI.

Sperm candles have now been almost entirely superseded by the paraffin and stearine combination, but are still used to a small extent by gas examiners for light measurements ("standard candles").

The standard candle is allowed to contain 3—4·5 per cent. of beeswax, for overcoming the brittleness of the wax, and the spermaceti employed

[1] Scheithauer, *Die Fabrikation der Mineralöle*, Braunschweig, 1895.

should have a M.P. of 45—46° C. The candles should be examined for M.P., sapon. value, unsaponifiable, and acid value, the presence of paraffin wax and "stearine" being readily revealed respectively by the latter tests.

Example 1. A "paraffin" candle
The acid value was 30·5.
This showed the presence of about 15 per cent. of stearine, which was quite usual.
The candle is *unsuitable for a hot climate.*

Example 2. A wax candle for church use
This was stamped " beeswax, 75 per cent."
It gave the following figures :

Saponification value	...	...	57·0
Acid value	...	...	12·3
Unsaponifiable		...	73·5

The s.v. and the a.v., assuming mineral wax only were present, showed 60—62 per cent. beeswax only.
The unsaponifiable matter proved the presence of 30—40 per cent. of unsaponifiable substances in addition to that derived from the beeswax.
It was therefore reported as 13—15 *per cent. deficient in beeswax.*

Example 3. A "standard" sperm candle
This gave results as under :

Saponification value	...	...	12·0
Acid value	...	...	3·2
Ratio number	...	...	36·0

The INFERENCE from these figures is that the candle contained from 10—15 *per cent. of beeswax,* which is 5·5—9·5 per cent. in excess of the amount allowed.

E. Illuminating Oils

For pure mineral lubricating oils see page 217.
Saponifiable oils are still used to a large extent for illumination in lamps, as these keep alight in exposed situations when mineral oil lamps would become extinguished.
The chief oils employed are :

> Rape (colza) oil,
> Lard oil,
> Seal „
> Whale „
> Olive „

Semi-drying and drying oils are unsuitable owing to the rapid clogging of the wick. Mucilaginous substances, such as occur in imperfectly refined oils, are objectionable, and the free fatty acids should not exceed 5 per cent., otherwise charring of the wick is produced.
The varieties of rape known as *Jamba* and *Ravison* are not so suitable for burning as the genuine oil.
Mixtures of fatty with mineral oils are very common, and should be examined as under *lubricating oils* (page 217).

Example 1. A "lighthouse" oil

The oil burnt in a trial lamp for 15 hours without the wick clogging.

The oil was saponified with alcoholic potash, the alcohol evaporated off, and the soap dissolved in water.

It was then decomposed with dilute mineral acid.

The "middle layer" between the liquefied fatty acids and the aqueous liquor was examined for mucilaginous impurities. There was scarcely a trace of flocculent matter.

The iodine value was 135.

Free fatty acids (as oleic) 3·5 per cent.

Unsaponifiable matter 0·75 per cent.

Insoluble bromide value 18·5.

Bromides darkened at 200° C., but did not melt.

Cottonseed oil absent (by Halphen test).

INFERENCE. *A good quality seal oil, or seal-whale mixture, suitable for the purpose required.*

Example 2. A sample of "colza oil" for railway use

This gave figures as follows :

Free acids as oleic	...	2·3 per cent.
Saponification value	...	179
Titer test	... ...	18° C.
Valenta (true) $\dfrac{V \times 10}{80}$	...	11·2

From these figures adulteration with cottonseed oil was suspected.

The Halphen reaction proved cottonseed present.

From the above figures about 20 per cent. was inferred.

The oil was returned as *adulterated with cottonseed oil and unsuitable for the purpose required.*

F. Wax Mixtures and Polishes

Mixtures of waxes are used for various purposes, e.g. alone in "heel-ball" (gloss for leather soles of boots, etc.), photograph cylinders, dolls and toys, etc. and in the form of pastes and emulsions with turpentine and its substitutes in boot and shoe polishes, furniture and floor polishes, etc.

The analysis of a mixture of waxes, especially when coloured by aniline dyes, is a matter of considerable difficulty, and a good deal of reliance must be placed upon practical experience, and upon tests for hardness, brittleness, etc.

The most important guide is the melting point, and the saponification value and percentage unsaponifiable give a measure of the probable amount of mineral waxes present. (See special precautions for s.v. of waxes, page 106.)

The turpentine or other solvent may be obtained from shoe polishes, etc. by distillation in steam. The amount present in turpentine polishes varies from 60—80 per cent. It should be examined as under turpentine (p. 225).

The residual waxes may then be tested for M.P., etc. after being dried at 110° C. to constant weight.

The amount of *carbon-black* in black boot polishes may be obtained by the hot filtration of a turpentine solution of the polish. The aniline dye present is best estimated by colorimetric means (using standard solutions of nigrosene dyes in alcohol) in the filtrate from the above. The waxes employed for polishes are paraffin, ceresin, montan (crude and refined), carnaüba, candelilla, cane sugar, beeswax; and a few rarer varieties.

Some polishes are made entirely without turpentine, being jelly emulsions produced by partially saponifying waxes with alkali solutions, or by emulsifying them with soap solutions. Turpentine or petroleum distillates and water may both be present in a "saponified" polish.

Example. Sample of floor polish

Colour red, in tin box.

Steam distilled 100 c.c. of melted polish, this gave a yield of 65·6 c.c. liquid of S.G. 0·866.

Distilled 50 c.c.: 95 per cent. passed over below 180° C.

INFERENCE. *Pure turpentine.*

The residual wax gave the following figures:

Sapon. value	...	...	...	45·0
Acid value	...	...	...	10·2
Unsaponifiable		...	...	75·5
M.P.	...	...	...	52° C.

INFERENCE. *Probably a mixture of about equal proportions of beeswax and paraffin wax.*

SECTION IX

SCHEME FOR THE IDENTIFICATION OF AN OIL, FAT, OR WAX

(*Assuming no adulterant present*)

Note.—Although the following scheme may, it is confidently hoped, be found of service to the works' chemist, and to others connected with the oils' and kindred industries, it is designed primarily to familiarise students with the distinguishing features of the various subdivisions and classes of the large family of oily and waxy bodies.

The authors are of course fully aware that the distinctive appearance, or other external characteristic of an oil, fat, or wax—apart from outside information as to its source, etc.—will frequently give the key to its identity, or at any rate, to its class relationships, without the use of any scheme of detection.

Further, there are a few instances in which the natural variation of individual oils, etc. militates against hard and fast lines of demarcation (as e.g. with one or two vegetable fats). Cases of this kind are however rare, and will not interfere with the general value of a systematic scheme to the student.

Rarely-occurring oils and fats have been omitted, as their description is unlikely to be of use, at present, to technical men, and would only add an unnecessary complication.

SECTION IX

IDENTIFICATION OF A SIMPLE OIL, FAT, OR WAX

Note the PHYSICAL STATE of the substance.

If *liquid* (assuming a normal room temperature of 15—20° C.) it is an **oil** or **liquid wax**.

If *solid*, a **fat** or **wax**.

> The consistence may be soft and unctuous, and the fat is then often termed a "butter."

If an oil, it may be *fatty* ("natural") or *mineral*.

> The *liquid waxes* resemble fatty oils in appearance, and have usually a "fishy" odour, unless highly refined.

The natural oils and fats are **glycerides**, whilst the mineral oils and all waxes are **non-glycerides**. To distinguish, proceed as follows :

1. Place a small amount of the substance in a *perfectly dry* test-tube, and heat over a bunsen flame, with shaking, until distillation takes place or white fumes are evolved. A **penetrating odour** (of acroleïn) indicates a *glyceride*.

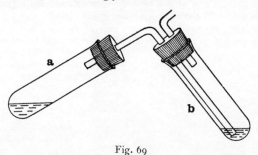

Fig. 69

2. Fit two small test-tubes ($5'' \times \frac{5}{8}''$) with corks and tubes as in fig. 69. Remove **a**, and place therein about 2 c.c. of the liquid, or 2 grms of the solid. Heat the test-tube **a** over a bunsen flame (cautiously at first) as in (1). Then attach to **b**, in which is placed a small

quantity of a 5 per cent. aqueous solution of *sodium nitroprusside* containing about 2 per cent. of *piperidine*. (The solution should reach *just* over the end of the delivery tube from **a**, so that liquid cannot be drawn back into **a**.) Heat for a few minutes strongly, until condensed oil appears on the surface of the liquid in **b**. In the case of a glyceride, acroleïn is evolved which turns the liquid in **b** a *bluish* or *heliotrope* colour. If the colour change is not distinctive, and, in any case to definitely confirm a *glyceride*, proceed as follows :

3. Take 20 grms of the substance, boil with alcoholic potash (as under Saponification Value, page 106), evaporate off alcohol, and acidify (using dilute sulphuric acid) the aqueous solution obtained by treating the residue with hot water, adding the acid till just pink to methyl orange. Heat on the water-bath till the oily layer on the surface is quite clear. Filter through a wet filter paper, and to the filtrate add a slight excess of barium carbonate to neutralise the excess of sulphuric acid. Filter again, and concentrate by heating on a water-bath in an evaporating basin. When crystallisation begins, allow to cool and treat with 95 per cent. alcohol, intermixing this with the crystals by means of a glass rod. Decant and filter (dry filter paper) into a 200 c.c. beaker. Wash again twice with alcohol, and evaporate the combined filtrates on the water-bath. A *syrupy residue* (of glycerin) is given with *glycerides*[1] (yield 1½—2½ grms). Non-glycerides yield no liquid residue.

Glycerides (natural oils and fats).

These vary very much in external character, especially as met with in commerce.

Note the **colour.**

This may vary from COLOURLESS to NEARLY BLACK.

OILS may be *ruby red* (as in case of crude cottonseed oil) or *green* (due to chlorophyll) in some vegetable oils. The most usual colour is *light yellow* to *brownish yellow*.

FATS vary usually from *white* to *brown*, according to the method of production, and the degree of refinement. Palm oil is orange to *orange red* in colour and some fats are *green*.

Note the **odour.**

Some highly refined or " cold pressed " oils and fats are almost free from odour. Many have strongly pronounced odours. Distinguish[2] as follows :

---

[1] The student may confirm by acetylation of the product (see page 188). For a *quantitative determination* of the glycerin see page 181.

[2] Although many odours, especially of oils and fats, have a fairly distinct and readily recognisable character, and are probably classified *mentally* by the expert, our *nomenclature of odours* is lamentably deficient and crude ; any attempt at classification is, on this account, a difficult matter.

QUALITY	DEGREE	SPECIFIC CHARACTER
"odourless"	apparently without odour	—
"pleasant" and "not unpleasant"	(1) "slight," or (2) "pronounced"	"nutty" "aromatic" "tallowy" "distinctive" (a recognised 　　or recalled odour) "unrecognisable"
"unpleasant" and "disagreeable"	(1) "slight" (2) "pronounced" (3) "acrid" (4) "penetrating" (5) "nauseous"	"fishy" "sulphurous" "goaty" "tallowy" "distinctive" "unrecognisable"

The **flavour** of an oil or fat varies greatly to the trained palate, but is, in general, distinctive only to the experts.

Note, if an oil, whether *clear* or *turbid*, or if there is a deposit of solid matter.

Oils vary very greatly in the amount of "stearine" or solid fat which they contain. Some, as e.g. linseed, almond, and castor oils, are rarely or never seen with deposited stearine at normal room temperatures. Of those which usually contain "stearine," the amount varies according to the temperature of the expression of the oil, and its subsequent treatment[1]. Castor oil is very viscous, and rape oil less so. Most other oils are fairly fluid at normal room temperatures.

Some fats are termed "oils" owing to their being liquid in the country of origin, e.g. coconut oil, palm kernel oil.

Two fats, "Japan wax" and "myrtle wax," have a waxy appearance and a conchoidal fracture on breaking. Hence their false description in commerce, these both being true glycerides.

Mineral oils (non-glycerides, liquid).

These are not so variable in character as the fatty oils, except as regards **consistence** and **viscosity**. Note if the consistence is *limpid, oily, thick* or *syrupy*.

Note the **colour.**

This may vary from *colourless* ("water white") to *dark reddish brown* or *black,* but the intermediate shades are almost invariably *reddish, yellowish red,* or *yellow.*

---

[1] Oil is often inadvertently decanted from solid residue.

Many mineral oils exhibit a *green fluorescence*, that is, they appear greenish by reflected light especially noticeable against a dark background.

Note the **odour.**

Most mineral oils have a recognisable odour, which in the lighter oils may be termed "*gassy*" (as somewhat reminiscent of coal gas) and in the heavier fractions "*empyreumatic*" (odour of a distilled product). Illuminating oils have usually the familiar odour of "paraffin oil."

Waxes (non-glycerides, solid).

A good deal of variation occurs between the waxes.

Note the **hardness.**

Some waxes are *very hard and brittle*, as e.g. carnaüba wax, insect wax and montan wax. Some of the mineral waxes are comparatively soft. *Plasticity* (capable of being "kneaded" with the fingers) is a characteristic quality of beeswax and ceresin (ozokerit) waxes.

Wool wax is exceptional in being *salve-like*, as also is vaseline.

Insect wax and spermaceti, also distilled montan wax, are *crystalline* and *fibrous* in structure.

On breaking most waxes, the broken surface has a *conchoidal* (shell-like) appearance.

Note the **colour.**

Beeswax (unless bleached) is *yellow* to *red* ; carnaüba wax, *yellowish* to *dirty green*. Ceresin is *yellow*, with (frequently) white specks; the paraffin waxes are usually *colourless* and *transparent* or *translucent*.

Note the **odour.**

Beeswax is *sweet* and *aromatic.* Carnaüba wax has, on melting a characteristic odour of *new-mown hay* ; wool wax has a pronounced "*goaty*" odour, while most of the other waxes are without smell.

Distinguish the waxes as follows:

Heat about 20 grms of the wax with alcoholic potash (as described, page 106), evaporate to dryness on water-bath, and take up with hot water. If the melted wax floats unaltered on the surface of a clear liquid, the wax is a MINERAL WAX.

If a soapy liquid is obtained, more or less cloudy (due to alcohols or hydrocarbons), the wax is a NATURAL (*animal* or *vegetable*) WAX, or a BITUMEN WAX.

The bitumen waxes are usually readily distinguished by their "earthy" odour, especially on warming.

A. GLYCERIDES

(1) LIQUIDS (natural oils)

Pour a few drops of the oil on the palm of one hand, rub the hands together for a few moments, and, keeping them almost closed, insert the nose in the space just over the thumbs, carefully noting the odour.

(i) A "fishy" odour, however faint, indicates **Marine oils (Class I).** "Deodorised" fish oils (so-called) may often be detected by proceeding as follows:

Saponify the oil with alcoholic potash (page 106), evaporate off the alcohol, dissolve the residue in water and bring to boil; acidulate with dilute sulphuric acid, and immediately note the odour of the liberated fatty acids. Under these conditions, almost all marine oils, even if highly refined, emit a distinct "fishy" odour.

"Hardened" marine oils, unless the hydrogenation has been carried very far, may be in this manner detected by the trained observer.

(ii) A not unpleasant, fresh, *distinctive* odour, sometimes markedly *aromatic* or *nutty*, points to a *vegetable* origin. **(Classes II, III, IV.)**

Ascertain the iodine value (page 92).

Iodine value **210** to about **120 Class II ("Drying oils").**

About   **120** „ „ **100 Class III ("Semi-drying oils").**

Below   **100**   **Class IV ("Non-drying oils").**

(iii) A faint *"tallowy"* odour indicates **Animal oils (Class V).** Since the only common animal oils, exclusive of the marine group, are *lard oil* and the *"foot oils,"* it should not be difficult to distinguish these from the vegetable oils.

If the odour is doubtful (and in any case to confirm) the unsaponifiable matter of the oil must be examined for *Cholesterol* or *Phytosterols* (see page 177).

(a) A *Phytosterol* is present. Vegetable oils (**Classes II, III, IV**). Confirm by the M.P. of the acetate (page 178).

(b) *Cholesterol* is present. Marine or animal oil (**Class I or V**). Ascertain the yield of *Insoluble Bromides* (page 109).

(i) Little or none. Animal oils (**Class V**).

(ii) Fair yield. Marine oils (**Class I**).

Confirm by ascertaining the melting points of the bromides produced.

The crystals (ii) should darken at 200° C. *without melting.*

CLASS I. MARINE OILS

(1) Dissolve about 2 c.c. of the oil in 10 c.c. CS_2 in a test-tube; add a few drops of concentrated sulphuric acid.

A *violet coloration* indicates LIVER OIL GROUP (GROUP 2).

(2) If the oil is light yellow in colour, and smells very pronouncedly of putrid fish, it probably belongs to the FISH OILS (GROUP 1).

(3) A large amount of "stearine," and, if light in colour, a mild fishy odour; or, if of a rank odour, then dark in colour, indicates BLUBBER OILS (GROUP 3).

Group 1. Fish Oils

These are extremely difficult to distinguish one from the other. The iodine values are very variable, but may give an indication.

(i) Iodine value above 150—probably *menhaden* or *salmon oil*.

SALMON OIL has reddish colour and mild odour. Average S.G. 924 at 15·5°.

MENHADEN usually brownish with rank odour. Average S.G. 931 at 15·5°.

(ii) Iodine value below 150—probably *Japan fish* or *Herring*. Ascertain insoluble bromide value (page 109).

40—50	...	...	...	...	JAPAN FISH.
10—25	...	...	...	...	HERRING.

Group 2. Liver Oils

Determine unsaponifiable matter (page 126).

Below 3—4 per cent.	...	...	COD LIVER OIL.
Above 4	„ „	...	... SHARK LIVER OIL.

Some shark oils contain a high proportion of the unsaturated hydrocarbon "*spinacene*[1]."

Other liver oils of commerce are generally described as "coast-cod" oils, and are seldom pure, but are mixtures of two or more liver oils. Discrimination is therefore out of the question.

Group 3. Blubber Oils

Ascertain the Reichert-Meissl value (page 113).

Below 5	...	...	...	WHALE OR SEAL OIL.
10—20	...	...	...	DOLPHIN BODY OIL.
20—40	...	...	...	PORPOISE BODY OIL.
90—120	...	...	...	DOLPHIN OR PORPOISE "JAW" OIL.

Seal oil has usually a higher iodine value and lower "titer" test (S.P. fatty acids) than whale oil. Dolphin and porpoise oils are comparatively rare.

CLASS II. VEGETABLE DRYING OILS

The iodine value already ascertained is the best guide.

205—206	...	...	...	PERILLA OIL.
170—200	...	...	...	LINSEED OIL.

Distinctive odour.

Confirm by high *insoluble bromide value* (page 112), (32—38).

[1] Chapman, *Analyst*, 1917, **42**, 161.

160—170   ...   . Tung oil.
Confirm by solidification test (page 133).

164—   ...   . Candlenut oil.
Distinguished from tung by insoluble bromide test.
Tung yields no bromides.
Candlenut gives 11—12 per cent.

148—158   ...   . Hempseed oil.
This oil is usually green in colour.

140—150   ...   . Walnut oil.
Characteristic odour of walnuts.

130—140   ...   . Poppyseed oil.
Distinguished from walnut, soya, niger and sunflower by yielding no insoluble bromides.

126—135   ...   . Soya bean oil.

126—133   ...   . Niger oil.

125—135   ...   . Sunflower oil.
The values of these three oils are very close and
they are distinguishable with difficulty by chemical
and physical tests. The odour and flavour are the
best guide. Sunflower oil is golden yellow in
colour, and niger oil has a distinct "nutty" flavour.

CLASS III. SEMI-DRYING OILS

The iodine value, as before, yields the best information.

115—125   ...   . Maize oil.
Confirm by distinctive odour and high unsaponifiable matter (due to lecithin).

111—120   ...   . Kapok oil.
Scarcely distinguishable from cottonseed, and gives
Halphen reaction, but *Becchi* test usually negative.

105—112   ...   . Cottonseed oil.
Confirm by Halphen and other colour tests and
by high s.p. fatty acids.

103—110   ...   . Sesamé oil.
Confirm by distinctive Baudouin colour test.

102—104   ...   . Croton oil.
Disagreeable odour and burning taste. Confirm
by Reichert-Meissl test = 12—13.

98—100   ...   . Curcas oil.
Unpleasant odour, but R.-M. value normal.

CLASS IV. NON-DRYING OILS

Ascertain saponification value (page 106) and confirm by *Valenta* test
(page 79).

s.v. 170—177. Rape oil group.   (Group I.)

176—184. CASTOR OIL GROUP. (GROUP 3.)
188—195. OLIVE OIL GROUP. (GROUP 2.)
Valenta (acetic acid).
Nearly insoluble boiling acetic. (Group 1.)
Soluble in cold acetic. (Group 3.)
Valenta normal (intermediate). (Group 2.)

Group 1. Rape Oil Group

The iodine values give an indication:

105—120 ... ... RAVISON OIL.
This is from a wild variety of rape seed.
105—110 ... ... MUSTARD OIL (black).
92—103 ... ... MUSTARD OIL (white).
Readily distinguished by characteristic pungent odour
(of ethereal mustard oil).
97—105 ... ... RAPE OIL.

Group 2. Olive Oil Group

The iodine values should be ascertained:

110—113 ... CHERRY KERNEL $\left.\right\}$
101—108 ... APRICOT „ $\left.\right\rbrace$ *The kernel oils.*
91—100 ... PLUM „
92—109 ... PEACH „
Readily distinguished from other oils of the group by
the "Bieber" reaction (colour test, p. 140).
98 100 ... ... ALMOND OIL.
The mixed fatty acids dissolve in an equal volume of
absolute alcohol at 15° C. to a clear solution.
(Distinct from arachis, rice, sesamé, etc.)
True Valenta no. $= 80 \cdot 0°$; $\dfrac{V \times 10}{80} = 10$ (p. 76).

91—97 ... ... RICE OIL.
Usually high acid value and distinctive "musty" odour.
88—95 ... ... ARACHIS OIL.
Distinguished by Bellier's and Renard's tests (pages
140—146).
True Valenta $= 88 \cdot 3°$; $\dfrac{V \times 10}{80} = 11 \cdot 0$ (p. 76).

80—85 ... ... OLIVE OIL.
Characteristic odour: test with liberated fatty acids.
True Valenta $= 72—76°$; $\dfrac{V \times 10}{80} = 9 \cdot 0—9 \cdot 5$ (p. 76).

18—2

Group 3. Castor Oil Group

The only commercial oil is CASTOR, readily distinguished by its

    (a) Solubility in alcohol.
    (b) High viscosity.
    (c) High specific gravity.
    (d) Acetyl value.

CLASS V. ANIMAL OILS

Iodine values as under:

    69—72 ... ... NEAT'S FOOT OIL.
    67—88 ... ... LARD OIL.
    The odour of lard in lard oil will usually serve to identify it.
    S.G. Neat's foot ·917 average.
    Lard oil ·9155 „

(2) SOLIDS. (The Fats)

The modern process of "hardening" oils into fats by hydrogenation with a metallic catalyst has made the identification of the source of fats as obtained in commerce a vastly more difficult problem than formerly. The recognition of hardened fats is dealt with on page 239.

In the case of naturally formed fats, there is usually little difficulty in distinguishing between vegetable fats (Class VI) and animal fats (Class VII) by means of the *odour* of the latter.   If this be very faint, the sample may be melted in a small basin, when a commercial sample of animal fat may in almost all cases be recognised with a little practice.

When doubt exists, and in all cases as confirmation, the isolation and identification of the *sterols* by means of the melting points of the acetates (p. 178) will give conclusive results.

CLASS VI. THE VEGETABLE FATS

Ascertain the *Saponification Value.*

    Below 200 ... GROUP 1 typical fat *Cacao butter.*
    200—225 ... GROUP 2   „   *Palm oil.*
    Above 225 ... GROUP 3   „   *Coconut oil.*

Group 1. The Vegetable Butters

Ascertain iodine value.

    75—85 ... ... LAUREL OIL.
    Recognised by green colour and characteristic odour.
    54—63 ... ... SHEA BUTTER.
    Has distinct odour, and high unsaponifiable (4—10 per cent.).

58—62 ... ... MOWRAH BUTTER.
(Latifolia fat.)
50—60 ... ... ILLIPÉ BUTTER.
(Longifolia fat.)
These two are often compounded and sold under either name.
The S.V. appears to be higher in Mowrah.
35—38 ... ... CACAO BUTTER.
Characterised by *Björklund's* test (page 146).
29—38 ... ... BORNEO TALLOW.
Resembles cacao butter closely.
Soluble baryta values[1] (page 155).
Cacao ... 2—7.
Borneo ... 8—15.

Group 2. Palm Oil Group. ("The Vegetable Tallows")

The iodine values are discriminative.
36—58 ... ... NUTMEG.
Distinguished very readily by its characteristic odour.
52—56 ... ... PALM OIL.
Orange to red colour, and characteristic odour.
(The latter persists on bleaching the fat.)
18—32 ... ... CHINESE VEGETABLE
TALLOW.
Titer ... 52°—53°.
Palm ... 36°—45°.
Japan wax ... 58°—60°.
4·5—15 ... ... JAPAN WAX.
Conchoidal fracture and distinctive odour.
M.P. 51°—53° C.
2—3 ... ... MYRTLE WAX.
Green colour from berries.
M.P. 40°—44° C.

Group 3. Coconut Oil Group

Characterised by glycerides of low molecular weight.
Iodine values.
10—17 ... ... PALM KERNEL.
8—9 ... ... COCONUT.
These fats are very similar. Palm kernel is rather variable.

	Coconut	Palm kernel
Polenske (p. 116)	17	11
True Valenta[2] $\frac{V \times 10}{80}$ (p. 76)...	1·5	3
Burnett and Revis test (p. 150)	52·5°	68·5°

[1] Bolton and Revis.
[2] Fryer and Weston.

CLASS VII. ANIMAL FATS

Ascertain the saponification value.
        Below 200   BODY FATS—GROUP 1.
        Above 200   MILK FATS—GROUP 2.

Group 1. Body Fats

Ascertain iodine value.
        Above 80   ...   Fish eating mammals (exclusive of marine
                        mammals, Class I, Group 3), ice-bear, etc.
        56—66   ...   .   LARD.
                        Readily identified by its translucent appearance and
                        salve-like consistence.
        50—55   ...   .   BONE FAT.
                        Recognised by presence of *lime salts.*
        38—45   ...   .   TALLOW.
                        Mutton tallow is commonly harder and stronger in
                        flavour and odour than beef fat.

Group 2. Milk Fats

The only common member is BUTTERFAT.
        Reichert-Meissl value (p. 113)   ...   25—30.
        Polenske (p. 116) ...   .   1·7—2·9.
        Kirschner (p. 154)   ...   ...   19—26.

B. NON-GLYCERIDES.   (Waxes and Mineral Oils)

Ascertain the saponification values.
    1. LIQUID AND SAPONIFIABLE   **Class VIII. Liquid waxes[1].**
    2. LIQUID AND UNSAPONIFIABLE   **„   XI.   Mineral oils.**
    3. SOLID AND SAPONIFIABLE   **Classes IX and X. (Natural
                                    waxes and bitumen waxes.)**
    4. SOLID AND UNSAPONIFIABLE   **Class XII. Mineral waxes.**

CLASS VIII. LIQUID WAXES

Two only are of importance.
Ascertain specific gravity.
        ·878—·883   ...   .   SPERM OIL.
        ·878—·880   ...   .   ARCTIC SPERM OIL.

---

[1] Certain oils distilled from rosin (rosin and pine oils) are slightly saponifiable
owing to the rosin acids they contain : they are detected by the Liebermann-Storch
reaction (page 222).

CLASS IX. NATURAL WAXES

It is not at present possible to ascertain whether a wax has an animal or vegetable origin. It is therefore assumed that the source is unknown. Ascertain *melting point* (page 38).

83·5°—84·5° ... ... CARNAÜBA WAX.
Confirm by S.G. = ·999 at 15·5° C.
Melt small quantity—agreeable odour of new-mown hay.
Turbidity temp.[1] alcohol reagent = 82° C.

80°—82° ... ... INSECT WAX.
Easily recognised by its fibrous and lustrous appearance. Resembles spermaceti but is much harder.
Turbidity temp.[1] alcohol reagent = insoluble.

68°—70° ... ... CANDELILLA WAX.
Characteristic odour.
Turbidity temp.[1] alcohol reagent = 63° C.

62°—65° ... ... BEESWAX.
Usually brownish yellow, with pleasant odour of honey.
Turbidity temp.[1] alcohol reagent = 76° C.
"Ratio number" = 3·6—3·8.

42°—49° ... ... SPERMACETI.
Lustrous translucent scales, almost odourless.
Turbidity temp.[1] alcohol reagent = 44° C.

30°—40° ... ... WOOL WAX.
Recognised by its "goaty" odour, salve-like consistence, and ready absorption of water.

CLASS X. BITUMEN WAXES

CRUDE BITUMEN WAX.
Recognised by its lustrous, brownish black colour, highly brittle character, and bituminous odour on warming.

REFINED BITUMEN WAX.
The acid refined product resembles ceresin wax closely, but can be distinguished therefrom by the production of a *fatty acid* (montanic acid) on distillation with superheated steam, also by its *acid value* = (15—20).

"MONTAN WAX."
A distillation product of bitumen wax; has fibrous crystalline structure, brown to white colour, and faint bituminous odour. Distinguished by its high acid value = 73—85.

[1] Fryer and Weston. See page 86.

PEAT WAX.

A distilled product of peat, almost indistinguishable from montan wax.

CLASS XI. MINERAL OILS

Note colour :

(1)   Very dark brown to black[1], with green fluorescence.
Fractionally distil 100 c.c. (page 214).

(*a*)   No liquid distils till high temperature
reached.   PETROLEUM RESIDUE.

(*b*)   Many fractions obtained, varying from
light to heavy.   CRUDE PETROLEUM.

For ascertaining source of oil, compare fractions obtained with those given in Vol. I.

(2)   The oil is *colourless* to *red brown.*

(*a*)   Ascertain specific gravity.

(i)   The oil is light and volatilises on hand.

·590—·668   ...   .   PETROLEUM ETHER.

·668—·747   ...   .   PETROLEUM SPIRIT.

(ii)   The oil is mobile, but not readily volatile, gradually
evaporating when dropped on filter paper.

·75—·80   ...   .   WHITE SPIRIT.

Flash point 75°—80° F. = turpentine substitute.

(iii)   A drop of oil placed on a piece of filter paper does
not evaporate.

·80—·82   ...   .   ILLUMINATING OIL.

·82—·84   ...   .   "PARAFFIN OIL."

·84—·86   ...   .   "SOLAR OIL."

·86—·92   ...   .   LUBRICATING OILS.

(*a*)   Fairly mobile, light colour.
"SPINDLE OIL."

(*β*)   Medium fluidity with strong
green fluorescence.   "ENGINE" OIL.

(*γ*)   Thick, flows very slowly,
darker colour.   CYLINDER OIL.

(*b*)   Cool the oil and note temperature of
setting.

(i)   Above 0° C.   ...   probably   SHALE OIL.

(ii)   Between 0° C. and − 10° C. „   AMERICAN OIL.

(iii)   Fluid at − 18° C.   „   RUSSIAN OIL.

[1] A few crude petroleums are lighter in colour.

CLASS XII. MINERAL WAXES

Ascertain melting points.

30°— 32° ... ... " VASELINE."

Of salve-like consistence, non-crystalline.

36°—60° ... ... PARAFFIN WAX.

Transparent—sometimes opaque : water white or slightly yellow. Crystalline.

Examine crystals from benzene under microscope.

Above 68° ... ... OZOKERITE (CERESIN).

Translucent light yellow colour. Plastic when warmed.

Non-crystalline.

SECTION X

TABLES

TABLE I.

Acid values : *in terms of oleic acid.*

Acid value	Oleic acid	Acid value	Oleic acid	Acid value	Oleic acid	Acid value	Oleic acid	Acid value	Oleic acid
1	0·50	41	20·61	81	40·72	121	60·83	161	80·93
2	1·01	42	21·11	82	41·22	122	61·33	162	81·44
3	1·51	43	21·62	83	41·73	123	61·83	163	81·94
4	2·01	44	22·12	84	42·23	124	62·33	164	82·44
5	2·51	45	22·62	85	42·73	125	62·84	165	82·94
6	3·02	46	23·12	86	43·23	126	63·34	166	83·45
7	3·52	47	23·63	87	43·74	127	63·84	167	83·95
8	4·02	48	24·13	88	44·24	128	64·34	168	84·45
9	4·52	49	24·63	89	44·74	129	64·85	169	84·95
10	5·03	50	25·14	90	45·24	130	65·35	170	85·46
11	5·53	51	25·64	91	45·75	131	65·85	171	85·96
12	6·03	52	26·14	92	46·25	132	66·35	172	86·46
13	6·53	53	26·64	93	46·75	133	66·86	173	86·96
14	7·04	54	27·15	94	47·25	134	67·36	174	87·47
15	7·51	55	27·65	95	47·76	135	67·86	175	87·97
16	8·04	56	28·15	96	48·26	136	68·36	176	88·47
17	8·55	57	28·65	97	48·76	137	68·87	177	88·97
18	9·05	58	29·16	98	49·26	138	69·37	178	89·48
19	9·55	59	29·56	99	49·77	139	69·87	179	89·98
20	10·05	60	30·16	100	50·27	140	70·38	180	90·48
21	10·56	61	30·66	101	50·77	141	70·88	181	90·98
22	11·06	62	31·17	102	51·28	142	71·38	182	91·49
23	11·56	63	31·67	103	51·78	143	71·88	183	91·99
24	12·06	64	32·17	104	52·28	144	72·39	184	92·49
25	12·57	65	32·67	105	52·78	145	72·89	185	92·99
26	13·07	66	33·18	106	53·28	146	73·39	186	93·50
27	13·57	67	33·68	107	53·79	147	73·89	187	94·00
28	14·07	68	34·18	108	54·29	148	74·40	188	94·50
29	14·58	69	34·68	109	54·79	149	74·90	189	95·00
30	15·08	70	35·19	110	55·30	150	75·41	190	95·51
31	15·58	71	35·69	111	55·80	151	75·91	191	96·01
32	16·08	72	36·19	112	56·30	152	76·41	192	96·51
33	16·59	73	36·69	113	56·80	153	76·91	193	97·01
34	17·09	74	37·20	114	57·31	154	77·42	194	97·52
35	17·59	75	37·70	115	57·81	155	77·92	195	98·02
36	18·10	76	38·20	116	58·31	156	78·42	196	98·52
37	18·60	77	38·70	117	58·81	157	78·92	197	99·02
38	19·10	78	39·21	118	59·32	158	79·43	198	99·52
39	19·60	79	39·71	119	59·82	159	79·93	198·9	100·00
40	20·11	80	40·22	120	60·32	160	80·43		

TABLE II.

Alcohol (*Ethyl*). S.G. *of solutions* (15·5° C.).

Per cent. alcohol	S.G. corresponding to		Per cent. alcohol	S.G. corresponding to	
	Per cent. by volume	Per cent. by weight		Per cent. by volume	Per cent. by weight
1	0·9985	0·9981	51	0·9323	0·9160
2	·9970	·9963	52	·9303	·9138
3	·9956	·9944	53	·9283	·9116
4	·9942	·9928	54	·9263	·9094
5	·9928	·9912	55	·9242	·9072
6	·9915	·9896	56	·9221	·9049
7	·9902	·9880	57	·9200	·9027
8	·9890	·9866	58	·9178	·9004
9	·9878	·9852	59	·9156	·8981
10	·9866	·9839	60	·9134	·8958
11	·9854	·9826	61	·9112	·8935
12	·9843	·9813	62	·9090	·8911
13	·9832	·9800	63	·9067	·8888
14	·9821	·9788	64	·9044	·8865
15	·9811	·9775	65	·9021	·8842
16	·9800	·9763	66	·8997	·8818
17	·9790	·9751	67	·8973	·8795
18	·9780	·9739	68	·8949	·8772
19	·9770	·9727	69	·8925	·8748
20	·9760	·9714	70	·8900	·8724
21	·9750	·9702	71	·8875	·8700
22	·9740	·9690	72	·8850	·8676
23	·9729	·9677	73	·8825	·8652
24	·9719	·9664	74	·8799	·8629
25	·9709	·9651	75	·8773	·8605
26	·9698	·9637	76	·8747	·8581
27	·9688	·9622	77	·8720	·8557
28	·9677	·9607	78	·8693	·8533
29	·9666	·9592	79	·8666	·8509
30	·9655	·9577	80	·8639	·8484
31	·9643	·9560	81	·8611	·8459
32	·9631	·9544	82	·8583	·8435
33	·9618	·9526	83	·8555	·8409
34	·9605	·9508	84	·8526	·8385
35	·9592	·9490	85	·8496	·8359
36	·9579	·9472	86	·8466	·8333
37	·9565	·9453	87	·8436	·8307
38	·9550	·9433	88	·8405	·8282
39	·9535	·9413	89	·8373	·8256
40	·9519	·9394	90	·8339	·8229
41	·9503	·9374	91	·8306	·8203
42	·9487	·9353	92	·8272	·8176
43	·9470	·9332	93	·8235	·8149
44	·9452	·9311	94	·8201	·8122
45	·9435	·9291	95	·8164	·8094
46	·9417	·9269	96	·8125	·8065
47	·9399	·9248	97	·8084	·8036
48	·9381	·9227	98	·8041	·8006
49	·9362	·9204	99	·7995	·7976
50	·9343	·9183	100	·7946	·7946

TABLE III.

Ammonia.

S.G. *of solutions* $\left(\dfrac{15^\circ}{4^\circ}\ C.\right)$ *and per cent. by weight of* NH_3.

S.G. $\frac{15^\circ}{4^\circ}$ C.	NH_3 per cent.	S.G. $\frac{15^\circ}{4^\circ}$ C.	NH_3 per cent.	S.G. $\frac{15^\circ}{4^\circ}$ C.	NH_3 per cent.
·996	0·91	·956	11·03	·916	23·03
·994	1·37	·954	11·59	·914	23·67
·992	1·84	·952	12·17	·912	24·33
·990	2·31	·950	12·74	·910	24·98
·988	2·80	·948	13·31	·908	25·65
·986	3·29	·946	13·88	·906	26·31
·984	3·80	·944	14·46	·904	26·98
·982	4·30	·942	15·04	·902	27·65
·980	4·80	·940	15·63	·900	28·33
·978	5·30	·938	16·22	·898	29·00
·976	5·80	·936	16·82	·896	29·69
·974	6·30	·934	17·42	·894	30·36
·972	6·80	·932	18·03	·892	31·05
·970	7·31	·930	18·63	·890	31·75
·968	7·82	·928	19·25	·888	32·50
·966	8·33	·926	19·86	·886	33·30
·964	8·84	·924	20·49	·884	34·10
·962	9·37	·922	21·11	·882	34·95
·960	9·91	·920	21·75	·880	35·70
·958	10·47	·918	22·38		

TABLE IV.

1918

International Atomic Weights.

	Symbol	Atomic weight		Symbol	Atomic weight
Aluminium	Al	27·1	Molybdenum	Mo	96·0
Antimony	Sb	120·2	Neodymium	Nd	144·3
Argon	A	39·88	Neon	Ne	20·2
Arsenic	As	74·96	Nickel	Ni	58·68
Barium	Ba	137·37	Niton (radium emanation)	Nt	222·4
Bismuth	Bi	208·0	Nitrogen	N	14·01
Boron	B	11·0	Osmium	Os	190·9
Bromine	Br	79·92	Oxygen	O	16·00
Cadmium	Cd	112·40	Palladium	Pd	106·7
Cæsium	Cs	132·81	Phosphorus	P	31·04
Calcium	Ca	40·07	Platinum	Pt	195·2
Carbon	C	12·005	Potassium	K	39·10
Cerium	Ce	140·25	Praseodymium	Pr	140·9
Chlorine	Cl	35·46	Radium	Ra	226·0
Chromium	Cr	52·0	Rhodium	Rh	102·9
Cobalt	Co	58·97	Rubidium	Rb	85·45
Columbium	Cb	93·5	Ruthenium	Ru	101·7
Copper	Cu	63·57	Samarium	Sa	150·4
Dysprosium	Dy	162·5	Scandium	Sc	44·1
Erbium	Er	167·7	Selenium	Se	79·2
Europium	Eu	152·0	Silicon	Si	28·3
Fluorine	F	19·0	Silver	Ag	107·88
Gadolinium	Gd	157·3	Sodium	Na	23·00
Gallium	Ga	69·9	Strontium	Sr	87·63
Germanium	Ge	72·5	Sulphur	S	32·06
Glucinum	Gl	9·1	Tantalum	Ta	181·5
Gold	Au	197·2	Tellurium	Te	127·5
Helium	He	4·00	Terbium	Tb	159·2
Holmium	Ho	163·5	Thallium	Tl	204·0
Hydrogen	H	1·008	Thorium	Th	232·4
Indium	In	114·8	Thulium	Tm	168·5
Iodine	I	126·92	Tin	Sn	118·7
Iridium	Ir	193·1	Titanium	Ti	48·1
Iron	Fe	55·84	Tungsten	W	184·0
Krypton	Kr	82·92	Uranium	U	238·2
Lanthanum	La	139·0	Vanadium	V	51·0
Lead	Pb	207·20	Xenon	Xe	130·2
Lithium	Li	6·94	Ytterbium (Neoytterbium)	Yb	173·5
Lutecium	Lu	175·0	Yttrium	Yt	88·7
Magnesium	Mg	24·32	Zinc	Zn	65·37
Manganese	Mn	54·93	Zirconium	Zr	90·6
Mercury	Hg	200·6			

TABLE V.

Capacities *of cylindrical vessels of* 1 *cm. length or height.*

The capacity of any vessel may be obtained by multiplying the figure
in c.c. by the length or height of the vessel in question (in cm.).

Diameter cm.	Capacity c.c.	Diameter cm.	Capacity c.c.	Diameter cm.	Capacity c.c.
1	0·7854	35	962·1127	68	3631·6811
2	3·1416	36	1017·8760	69	3739·2806
3	7·0686	37	1075·2101	70	3848·4510
4	12·5663	38	1134·1149	71	3959·1921
5	19·6349	39	1194·5906	72	4071·5641
6	28·2743	40	1256·6370	73	4185·3868
7	38·4845	41	1320·2543	74	4300·8403
8	50·2654	42	1385·4423	75	4417·8646
9	63·6172	43	1452·2012	76	4536·4598
10	78·5398	44	1520·5308	77	4656·6257
11	95·0332	45	1590·4313	78	4778·3624
12	113·0973	46	1661·9025	79	4901·6699
13	132·7323	47	1734·9445	80	5026·5482
14	153·9380	48	1809·5574	81	5152·9973
15	176·7146	49	1885·7410	82	5281·0172
16	201·0619	50	1963·4954	83	5410·6080
17	226·9807	51	2042·8206	84	5541·7694
18	254·4690	52	2123·7166	85	5674·5017
19	283·5287	53	2206·1834	86	5808·8048
20	314·1593	54	2290·2210	87	5944·6787
21	346·3606	55	2375·8294	88	6082·1233
22	380·1327	56	2463·0086	89	6221·1388
23	415·4756	57	2551·7586	90	6361·7251
24	452·3893	58	2642·3794	91	6503·8822
25	490·8738	59	2733·9710	92	6647·6100
26	530·9291	60	2827·4334	93	6792·9087
27	572·5553	61	2922·4665	94	6939·7781
28	615·7521	62	3019·0705	95	7088·2184
29	660·5198	63	3117·2453	96	7238·2294
30	706·8583	64	3216·9908	97	7389·8113
31	754·7676	65	3318·3072	98	7542·9639
32	804·2477	66	3421·1944	99	7697·6874
33	855·2986	67	3525·6523	100	7853·9816
34	907·9203				

Applicable also to any unit of length and its cube: inches—cu. inches;
feet—cu. feet; metres—cu. metres, etc.

TABLE VI. 289

Carbonate of potash. S.G. *of solutions* 15° C. *(Gerlach)*.

S.G.	Twaddell	Per cent. by weight. K_2CO_3	Kilogrm. per cubic metre. K_2CO_3	Lbs. per cubic foot. K_2CO_3	S.G.	Twaddell	Per cent. by weight. K_2CO_3	Kilogrm. per cubic metre. K_2CO_3	Lbs. per cubic foot. K_2CO_3
1·005	1	·54	5·4	0·34	1·290	58	29·02	374·3	23·34
1·010	2	1·08	10·9	0·68	1·295	59	29·46	381·5	23·79
1·015	3	1·62	16·4	1·02	1·300	60	29·91	388·8	24·24
1·020	4	2·16	22·0	1·37	1·305	61	30·34	395·9	24·68
1·025	5	2·70	27·7	1·73	1·310	62	30·77	403·1	25·13
1·030	6	3·24	33·4	2·08	1·315	63	31·21	410·3	25·58
1·035	7	3·78	39·1	2·43	1·320	64	31·64	417·6	26·04
1·040	8	4·32	44·9	2·80	1·325	65	32·08	425·0	26·50
1·045	9	4·86	50·8	3·17	1·330	66	32·51	432·4	26·96
1·050	10	5·40	56·7	3·53	1·335	67	32·94	439·8	27·42
1·055	11	5·94	62·7	3·90	1·340	68	33·38	447·3	27·89
1·060	12	6·48	68·7	4·28	1·345	69	33·81	454·8	28·36
1·065	13	7·02	74·8	4·66	1·350	70	34·25	462·4	28·83
1·070	14	7·56	80·9	5·04	1·355	71	34·67	469·9	29·30
1·075	15	8·10	87·1	5·43	1·360	72	35·10	477·4	29·77
1·080	16	8·64	93·3	5·82	1·365	73	35·52	484·9	30·23
1·085	17	9·18	99·6	6·21	1·370	74	35·95	492·5	30·71
1·090	18	9·72	105·9	6·60	1·375	75	36·37	500·1	31·18
1·095	19	10·26	108·4	6·51	1·380	76	36·80	507·8	31·66
1·100	20	10·80	118·8	7·41	1·385	77	37·22	515·6	32·15
1·105	21	11·31	125·0	7·79	1·390	78	37·65	523·3	32·63
1·110	22	11·82	131·2	8·18	1·395	79	38·07	531·7	33·11
1·115	23	12·33	137·5	8·57	1·400	80	38·50	539·0	33·60
1·120	24	12·84	143·8	8·97	1·405	81	38·91	546·7	34·09
1·125	25	13·35	150·2	9·37	1·410	82	39·32	554·4	34·57
1·130	26	13·86	156·6	9·76	1·415	83	39·73	562·2	35·05
1·135	27	14·37	163·1	10·17	1·420	84	40·14	570·0	35·54
1·140	28	14·88	169·6	10·57	1·425	85	40·55	577·8	36·02
1·145	29	15·39	176·2	10·99	1·430	86	40·96	585·7	36·51
1·150	30	15·90	182·8	11·40	1·435	87	41·37	593·6	37·01
1·155	31	16·38	189·2	11·80	1·440	88	41·78	601·6	37·51
1·160	32	16·86	195·6	12·20	1·445	89	42·19	609·6	38·01
1·165	33	17·34	202·0	12·59	1·450	90	42·60	617·7	38·51
1·170	34	17·82	208·5	13·00	1·455	91	43·00	625·6	39·01
1·175	35	18·30	215·0	13·40	1·460	92	43·40	633·6	39·51
1·180	36	18·78	221·6	13·82	1·465	93	43·80	641·6	40·01
1·185	37	19·26	228·2	14·23	1·470	94	44·20	649·7	40·51
1·190	38	19·74	234·9	14·65	1·475	95	44·60	657·8	41·01
1·195	39	20·22	241·7	15·07	1·480	96	45·00	666·0	41·52
1·200	40	20·70	248·4	15·49	1·485	97	45·40	674·2	42·03
1·205	41	21·17	255·2	15·91	1·490	98	45·80	682·4	42·55
1·210	42	21·65	262·0	16·33	1·495	99	46·20	690·7	43·06
1·215	43	22·12	268·8	16·76	1·500	100	46·60	699·0	43·58
1·220	44	22·60	275·7	17·19	1·505	101	46·98	707·1	44·09
1·225	45	23·07	282·6	17·62	1·510	102	47·37	715·3	44·61
1·230	46	23·55	289·6	18·05	1·515	103	47·75	723·5	45·11
1·235	47	24·02	296·7	18·50	1·520	104	48·14	731·7	45·62
1·240	48	24·50	303·8	18·94	1·525	105	48·52	740·0	46·14
1·245	49	24·97	310·9	19·38	1·530	106	48·91	748·3	46·66
1·250	50	25·45	318·1	19·83	1·535	107	49·29	756·7	47·18
1·255	51	25·89	325·0	20·26	1·540	108	49·68	765·1	47·70
1·260	52	26·34	331·9	20·70	1·545	109	50·06	773·5	48·22
1·265	53	26·78	338·8	21·12	1·550	110	50·45	782·0	48·76
1·270	54	27·23	345·8	21·56	1·555	111	50·83	790·5	49·29
1·275	55	27·68	352·8	22·00	1·560	112	51·22	799·0	49·82
1·280	56	28·12	359·9	22·44	1·565	113	51·61	807·7	50·36
1·285	57	28·57	367·1	22·89	1·570	114	52·00	816·4	50·90

TABLE VII.

Carbonate of soda.

S.G. *of solutions and per cent. by weight of* Na_2O *and* Na_2CO_3.

S.G. 15·5° C.	Degrees Twaddell	Na_2O	Na_2CO_3
1·005	1	0·28	0·47
1·010	2	0·56	0·95
1·015	3	0·84	1·42
1·020	4	1·11	1·90
1·025	5	1·39	2·38
1·030	6	1·67	2·85
1·035	7	1·95	3·33
1·040	8	2·22	3·80
1·045	9	2·50	4·28
1·050	10	2·78	4·76
1·055	11	3·06	5·23
1·060	12	3·34	5·71
1·065	13	3·61	6·17
1·070	14	3·88	6·64
1·075	15	4·16	7·10
1·080	16	4·42	7·57
1·085	17	4·70	8·04
1·090	18	4·97	8·51
1·095	19	5·24	8·97
1·100	20	5·52	9·43
1·105	21	5·79	9·90
1·110	22	6·06	10·37
1·115	23	6·33	10·83
1·120	24	6·61	11·30
1·125	25	6·88	11·76
1·130	26	7·15	12·23
1·135	27	7·42	12·70
1·140	28	7·70	13·16
1·145	29	7·97	13·63
1·150	30	8·24	14·09

Table VIII.

Caustic alkali. S.G. *of solutions* ($15 \cdot 5^\circ$ C.) *and per cent. by weight of* NaOH *and* KOH.

S.G.	NaOH. Per cent.	KOH. Per cent.	S.G.	NaOH. Per cent.	KOH. Per cent.
1·0035	0·30	0·45	1·2152	19·08	23·75
1·0070	0·61	0·90	1·2202	19·58	24·20
1·0105	0·90	1·30	1·2255	20·08	24·65
1·0141	1·20	1·70	1·2308	20·59	25·10
1·0177	1·60	2·15	1·2361	21·00	25·60
1·0213	2·00	2·60	1·2414	21·42	26·10
1·0249	2·36	3·05	1·2468	22·03	26·50
1·0286	2·71	3·50	1·2522	22·64	27·00
1·0323	3·03	4·00	1·2576	23·15	27·50
1·0360	3·35	4·50	1·2632	23·67	28·00
1·0397	3·67	5·05	1·2687	24·24	28·45
1·0435	4·00	5·60	1·2743	24·81	28·90
1·0473	4·32	6·00	1·2800	25·30	29·35
1·0511	4·64	6·40	1·2857	25·80	29·80
1·0549	4·96	6·80	1·2905	26·31	30·25
1·0588	5·29	7·40	1·2973	26·83	30·70
1·0627	5·58	7·80	1·3032	27·31	31·25
1·0667	5·87	8·20	1·3091	27·80	31·80
1·0706	6·21	8·70	1·3151	28·31	32·25
1·0746	6·55	9·20	1·3211	28·83	32·70
1·0787	6·76	9·65	1·3272	29·38	33·20
1·0827	7·31	10·10	1·3333	29·93	33·70
1·0868	7·66	10·50	1·3395	30·57	34·30
1·0909	8·00	10·90	1·3458	31·22	34·90
1·0951	8·34	11·45	1·3521	31·85	35·40
1·1000	8·68	12·00	1·3585	32·47	35·90
1·1035	9·05	12·45	1·3649	33·08	36·40
1·1077	9·42	12·90	1·3714	33·69	36·90
1·1120	9·74	13·35	1·3780	34·38	37·35
1·1163	10·06	13·80	1·3846	34·96	37·80
1·1206	10·51	14·30	1·3913	35·65	38·35
1·1250	10·97	14·80	1·3981	36·25	38·90
1·1294	11·42	15·25	1·4049	36·86	39·40
1·1339	11·84	15·70	1·4187	38·13	40·40
1·1383	12·24	16·10	1·4267	38·80	40·90
1·1423	12·64	16·50	1·4328	39·39	41·50
1·1474	13·00	17·15	1·4400	39·99	42·10
1·1520	13·55	17·60	1·4472	40·75	42·75
1·1566	13·86	18·10	1·4545	41·41	43·40
1·1613	14·37	18·60	1·4619	42·12	44·00
1·1660	14·75	19·05	1·4694	42·83	44·60
1·1707	15·13	19·50	1·4769	43·66	45·20
1·1755	15·50	20·00	1·4845	44·38	45·80
1·1803	15·91	20·50	1·4922	45·27	46·45
1·1852	16·38	20·95	1·5000	46·15	47·10
1·1901	16·77	21·40	1·5079	46·87	47·70
1·1950	17·22	21·90	1·5158	47·60	48·30
1·2000	17·67	22·50	1·5238	48·81	48·85
1·2050	18·12	22·85	1·5319	49·02	49·40
1·2101	18·58	23·30			

TABLE IX.

Centigrade degrees equivalent to Fahrenheit.

Fahr.	Cent.	Fahr.	Cent.	Fahr.	Cent.	Fahr.	Cent.
0	− 17·8	54	+ 12·2	107	+41·7	160	+71·1
1	17·2	55	12·8	108	42·2	161	71·7
2	16·7	56	13·3	109	42·8	162	72·2
3	16·1	57	13·9	110	43·3	163	72·8
4	15·6	58	14·4	111	43·9	164	73·3
5	15·0	59	15·0	112	44·4	165	73·9
6	14·4	60	15·6	113	45·0	166	74·4
7	13·9	61	16·1	114	45·6	167	75·0
8	13·3	62	16·7	115	46·1	168	75·6
9	12·8	63	17·2	116	46·7	169	76·1
10	12·2	64	17·8	117	47·2	170	76·7
11	11·7	65	18·3	118	47·8	171	77·2
12	11·1	66	18·9	119	48·3	172	77·8
13	10·6	67	19·4	120	48·9	173	78·3
14	10·0	68	20·0	121	49·4	174	78·9
15	9·4	69	20·6	122	50·0	175	79·4
16	8·9	70	21·1	123	50·6	176	80·0
17	8·3	71	21·7	124	51·1	177	80·6
18	7·8	72	22·2	125	51·7	178	81·1
19	7·2	73	22·8	126	52·2	179	81·7
20	6·7	74	23·3	127	52·8	180	82·2
21	6·1	75	23·9	128	53·3	181	82·8
22	5·6	76	24·4	129	53·9	182	83·3
23	5·0	77	25·0	130	54·4	183	83·9
24	4·4	78	25·6	131	55·0	184	84·4
25	3·9	79	26·1	132	55·6	185	85·0
26	3·3	80	26·7	133	56·1	186	85·6
27	2·8	81	27·2	134	56·7	187	86·1
28	2·2	82	27·8	135	57·2	188	86·7
29	1·7	83	28·3	136	57·8	189	87·2
30	1·1	84	28·9	137	58·3	190	87·8
31	0·6	85	29·4	138	58·9	191	88·3
32	+0·0	86	30·0	139	59·4	192	88·9
33	0·6	87	30·6	140	60·0	193	89·4
34	1·1	88	31·1	141	60·6	194	90·0
35	1·7	89	31·7	142	61·1	195	90·6
36	2·2	90	32·2	143	61·7	196	91·1
37	2·8	91	32·8	144	62·2	197	91·7
38	3·3	92	33·3	145	62·8	198	92·2
39	3·9	93	33·9	146	63·3	199	92·8
40	4·4	94	34·4	147	63·9	200	93·3
41	5·0	95	35·0	148	64·4	201	93·9
42	5·6	96	35·6	149	65·0	202	94·4
43	6·1	97	36·1	150	65·6	203	95·0
44	6·7	98	36·7	151	66·1	204	95·6
45	7·2	99	37·2	152	66·7	205	96·1
46	7·8	100	37·8	153	67·2	206	96·7
47	8·3	101	38·3	154	67·8	207	97·2
48	8·9	102	38·9	155	68·3	208	97·8
49	9·4	103	39·4	156	68·9	209	98·3
50	10·0	104	40·0	157	69·4	210	98·9
51	10·6	105	40·6	158	70·0	211	99·4
52	11·1	106	41·1	159	70·6	212	100·0
53	11·7						

TABLE X.

Fahrenheit degrees equivalent to Centigrade.

Cent.	Fahr.	Cent.	Fahr.	Cent.	Fahr.
−40	−40·0	+7	+44·6	+54	+129·2
39	38·2	8	46·4	55	131·0
38	36·4	9	48·2	56	132·8
37	34·6	10	50·0	57	134·6
36	32·8	11	51·8	58	136·4
35	31·0	12	53·6	59	138·2
34	29·2	13	55·4	60	140·0
33	27·4	14	57·2	61	141·8
32	25·6	15	59·0	62	143·6
31	23·8	16	60·8	63	145·4
30	22·0	17	62·6	64	147·2
29	20·2	18	64·4	65	149·0
28	18·4	19	66·2	66	150·8
27	16·6	20	68·0	67	152·6
26	14·8	21	69·8	68	154·4
25	13·0	22	71·6	69	156·2
24	11·2	23	73·4	70	158·0
23	9·4	24	75·2	71	159·8
22	7·6	25	77·0	72	161·6
21	5·8	26	78·8	73	163·4
20	4·0	27	80·6	74	165·2
19	2·2	28	82·4	75	167·0
18	0·4	29	84·2	76	168·8
17	+1·4	30	86·0	77	170·6
16	3·2	31	87·8	78	172·4
15	5·0	32	89·6	79	174·2
14	6·8	33	91·4	80	176·0
13	8·6	34	93·2	81	177·8
12	10·4	35	95·0	82	179·6
11	12·2	36	96·8	83	181·4
10	14·0	37	98·6	84	183·2
9	15·8	38	100·4	85	185·0
8	17·6	39	102·2	86	186·8
7	19·4	40	104·0	87	188·6
6	21·2	41	105·8	88	190·4
5	23·0	42	107·6	89	192·2
4	24·8	43	109·4	90	194·0
3	26·6	44	111·2	91	195·8
2	28·4	45	113·0	92	197·6
1	30·2	46	114·8	93	199·4
0	32·0	47	116·6	94	201·2
+1	33·8	48	118·4	95	203·0
2	35·6	49	120·2	96	204·8
3	37·4	50	122·0	97	206·6
4	39·2	51	123·8	98	208·4
5	41·0	52	125·6	99	210·2
6	42·8	53	127·4	100	212·0

TABLE XI.

Glycerin—Refractive indices and specific gravities (15° C.) *Skalweit*

Glycerol per cent.	S.G. at 15°C.	$n_{[D]}$ at 15°C.	Glycerol per cent.	S.G. at 15°C.	$n_{[D]}$ at 15°C.	Glycerol per cent.	S.G. at 15°C.	$n_{[D]}$ at 15°C.
0	1·0000	1·3330	34	1·0858	1·3771	68	1·1799	1·4265
1	1·0024	1·3342	35	1·0885	1·3785	69	1·1827	1·4280
2	1·0048	1·3354	36	1·0912	1·3799	70	1·1855	1·4295
3	1·0072	1·3366	37	1·0939	1·3813	71	1·1882	1·4309
4	1·0096	1·3378	38	1·0966	1·3827	72	1·1909	1·4324
5	1·0120	1·3390	39	1·0993	1·3840	73	1·1936	1·4339
6	1·0144	1·3402	40	1·1020	1·3854	74	1·1963	1·4354
7	1·0168	1·3414	41	1·1047	1·3868	75	1·1990	1·4369
8	1·0192	1·3426	42	1·1074	1·3882	76	1·2017	1·4384
9	1·0216	1·3439	43	1·1101	1·3896	77	1·2044	1·4399
10	1·0240	1·3452	44	1·1128	1·3910	78	1·2071	1·4414
11	1·0265	1·3464	45	1·1155	1·3924	79	1·2098	1·4429
12	1·0290	1·3477	46	1·1182	1·3938	80	1·2125	1·4444
13	1·0315	1·3490	47	1·1209	1·3952	81	1·2152	1·4460
14	1·0340	1·3503	48	1·1236	1·3966	82	1·2179	1·4475
15	1·0365	1·3516	49	1·1263	1·3981	83	1·2206	1·4490
16	1·0390	1·3529	50	1·1290	1·3996	84	1·2233	1·4505
17	1·0415	1·3542	51	1·1318	1·4010	85	1·2260	1·4520
18	1·0440	1·3555	52	1·1346	1·4024	86	1·2287	1·4535
19	1·0465	1·3568	53	1·1374	1·4039	87	1·2314	1·4550
20	1·0490	1·3581	54	1·1402	1·4054	88	1·2341	1·4565
21	1·0516	1·3594	55	1·1430	1·4069	89	1·2368	1·4580
22	1·0542	1·3607	56	1·1458	1·4084	90	1·2395	1·4595
23	1·0568	1·3620	57	1·1486	1·4099	91	1·2421	1·4610
24	1·0594	1·3633	58	1·1514	1·4104	92	1·2447	1·4625
25	1·0620	1·3647	59	1·1542	1·4129	93	1·2473	1·4640
26	1·0646	1·3660	60	1·1570	1·4144	94	1·2499	1·4655
27	1·0672	1·3674	61	1·1599	1·4160	95	1·2525	1·4670
28	1·0698	1·3687	62	1·1628	1·4175	96	1·2550	1·4684
29	1·0724	1·3701	63	1·1657	1·4190	97	1·2575	1·4698
30	1·0750	1·3715	64	1·1686	1·4205	98	1·2600	1·4712
31	1·0777	1·3729	65	1·1715	1·4220	99	1·2625	1·4728
32	1·0804	1·3743	66	1·1743	1·4235	100	1·2650	1·4742
33	1·0831	1·3757	67	1·1771	1·4250			

Correction for temperature.

S.G.	Variation of Refractive Index for 1° C.	Observer
1·25350	0·00032	Listing
1·24049	0·00025	Van der Willigen
1·19286	0·00023	,,
1·16270	0·00022	,,
1·11463	0·00021	,,

Table XII.

Glycerol.

Theoretical yields from neutral oils and fats (average).

Oil or fat	Glycerol per cent.	Oil or fat	Glycerol per cent.
Almond	10·4	Mustard	9·4
Apricot kernel	10·4	Myrtle wax	11·6
Arachis	10·5	Neat's foot	10·6
Bonefat	10·4	Niger	10·4
Borneo tallow	10·6	Olive	10·4
Butterfat	12·4	Palm	11·0
Cacao butter	10·6	Palm kernel	13·5
Candlenut	10·5	Peach kernel	10·4
Castor	9·8	Perilla	10·3
Chinese vegetable tallow	11·1	Poppyseed	10·5
Coconut	14·1	Rape	9·5
Cod liver	10·1	Rice	10·5
Cottonseed	10·6	Salmon	10·3
Hempseed	10·4	Seal	10·2
Herring	10·3	Sesamé	10·5
Illipé	10·3	Shark liver	1·1—10·2
Japan fish	10·4	Shea	10·2
Japan wax	12·0	Soya	10·5
Lard	10·7	Sunflower	10·4
Linseed	10·4	Tallow	10·7
Maize	10·4	Tung	10·4
Menhaden	10·5	Walnut	10·5
Mowrah	10·4	Whale	10·2

[To find the glycerin content in cases of free acidity of the oils, etc., it is approximately correct to deduct 0·1 from the glycerin figures for each one per cent. acidity as oleic acid.]

TABLE XIII.

Hydrochloric acid. S.G. *of solutions per cent. by weight of pure* HCl *and of concentrated solutions.*

S.G. $\frac{15°}{4}$	°Baumé	°Twaddell	HCl %	% S.G. 1·1425 (18° Bé.)	% S.G. 1·152 (19° Bé.)	% S.G. 1·163 (20° Bé.)	% S.G. 1·171 (21° Bé.)	% S.G. 1·180 (22° Bé.)
1·000	0·0	0·0	0·16	0·57	0·53	0·49	0·47	0·45
1·005	0·7	1	1·15	4·08	3·84	3·58	3·42	3·25
1·010	1·4	2	2·14	7·60	7·14	6·66	6·36	6·04
1·015	2·1	3	3·12	11·08	10·41	9·71	9·27	8·81
1·020	2·7	4	4·13	14·67	13·79	12·86	12·27	11·67
1·025	3·4	5	5·15	18·30	17·19	16·04	15·30	14·55
1·030	4·1	6	6·15	21·85	20·53	19·16	18·27	17·38
1·035	4·7	7	7·15	25·40	23·87	22·27	21·25	20·20
1·040	5·4	8	8·16	28·99	27·24	25·42	24·25	23·06
1·045	6·0	9	9·16	32·55	30·58	28·53	27·22	25·88
1·050	6·7	10	10·17	36·14	33·95	31·68	30·22	28·74
1·055	7·4	11	11·18	39·73	37·33	34·82	33·22	31·59
1·060	8·0	12	12·19	43·32	40·70	37·97	36·23	34·44
1·065	8·7	13	13·19	46·87	44·04	41·09	39·20	37·27
1·070	9·4	14	14·17	50·35	47·31	44·14	42·11	40·04
1·075	10·0	15	15·16	53·87	50·62	47·22	45·05	42·84
1·080	10·6	16	16·15	57·39	53·92	50·31	47·99	45·63
1·085	11·2	17	17·13	60·87	57·19	53·36	50·90	48·40
1·090	11·9	18	18·11	64·35	60·47	56·41	53·82	51·17
1·095	12·4	19	19·06	67·73	63·64	59·37	56·64	53·86
1·100	13·0	20	20·01	71·11	66·81	62·33	59·46	56·54
1·105	13·6	21	20·97	74·52	70·01	65·32	62·32	59·26
1·110	14·2	22	21·92	77·89	73·19	68·28	65·14	61·94
1·115	14·9	23	22·86	81·23	76·32	71·21	67·93	64·60
1·120	15·4	24	23·82	84·64	79·53	74·20	70·79	67·31
1·125	16·0	25	24·78	88·06	82·74	77·19	73·74	70·02
1·130	16·5	26	25·75	91·50	85·97	80·21	76·52	72·76
1·135	17·1	27	26·70	94·88	89·15	83·18	79·34	75·45
1·140	17·7	28	27·66	98·29	92·35	86·17	82·20	78·16
1·1425	18·0	28·5	28·14	100·00	93·95	87·66	83·62	79·51
1·145	18·3	29	28·61	101·67	95·52	89·13	85·02	80·84
1·150	18·8	30	29·57	105·08	98·73	92·11	87·87	83·55
1·152	19·0	30·4	29·95	106·43	100·00	93·30	89·01	84·63
1·155	19·3	31	30·55	108·58	102·00	95·17	90·79	86·32
1·160	19·8	32	31·52	112·01	105·24	98·19	93·67	89·07
1·163	20·0	32·6	32·10	114·07	107·17	100·00	95·39	90·70
1·165	20·3	33	32·49	115·46	108·48	101·21	96·55	91·81
1·170	20·9	34	33·46	118·91	111·71	104·24	99·43	94·55
1·171	21·0	34·2	33·65	119·58	112·35	104·82	100·00	95·09
1·175	21·4	35	34·42	122·32	114·92	107·22	102·28	97·26
1·180	22·0	36	35·39	125·76	118·16	110·24	105·17	100·00
1·185	22·5	37	36·31	129·03	121·23	113·11	107·90	102·60
1·190	23·0	38	37·23	132·30	124·30	115·98	110·63	105·20
1·195	23·5	39	38·16	135·61	127·41	118·87	113·40	107·83
1·200	24·0	40	39·11	138·98	130·58	121·84	116·22	110·51

TABLE XIV.

Nitric acid.

S.G. *of solutions* $\left(\dfrac{15^\circ}{4^c}\ C.\right)$ *and per cent. by weight of* HNO_3.

S.G. $\frac{15^\circ}{4^\circ}$	Twaddell	Baumé	HNO_3 per cent.	S.G. $\frac{15^\circ}{4^\circ}$	Twaddell	Baumé	HNO_3 per cent.
1·02	4	2·7	3·70	1·29	58	32·4	45·9
1·03	6	4·1	5·48	1·30	60	33·3	47·5
1·04	8	5·4	7·26	1·31	62	34·2	49·1
1·05	10	6·7	9·03	1·32	64	35·0	50·7
1·06	12	8·0	10·7	1·33	66	35·8	52·4
1·07	14	9·4	12·3	1·34	68	36·6	54·1
1·08	16	10·6	13·9	1·35	70	37·4	55·8
1·09	18	11·9	15·5	1·36	72	38·2	57·6
1·10	20	13·0	17·1	1·37	74	39·0	59·4
1·11	22	14·2	18·7	1·38	76	39·8	61·3
1·12	24	15·4	20·2	1·39	78	40·5	63·3
1·13	26	16·5	21·8	1·40	80	41·2	65·3
1·14	28	17·7	23·3	1·41	82	42·0	67·5
1·15	30	18·8	24·8	1·42	84	42·7	69·8
1·16	32	19·8	26·4	1·43	86	43·4	72·3
1·17	34	20·9	27·9	1·44	88	44·1	74·7
1·18	36	22·0	29·4	1·45	90	44·8	77·3
1·19	38	23·0	30·9	1·46	92	45·4	80·0
1·20	40	24·0	32·4	1·47	94	46·1	83·0
1·21	42	25·0	33·8	1·48	96	46·8	86·0
1·22	44	26·0	35·3	1·49	98	47·4	90·0
1·23	46	26·9	36·8	1·50	100	48·1	94·1
1·24	48	27·9	38·3	1·504	100·8	48·3	96·0
1·25	50	28·8	39·8	1·508	101·6	48·5	97·5
1·26	52	29·7	41·3	1·512	102·4	48·8	98·5
1·27	54	30·6	42·9	1·516	103·2	49·1	99·2
1·28	56	31·5	44·4	1·520	104·0	49·4	99·7

TABLE XV.

Oleic and Stearic acids. *Iodine values of mixtures.*

Iodine value	Oleic acid per cent.	Stearic acid per cent.	Iodine value	Oleic acid per cent.	Stearic acid per cent.
0	0	100	46	51·06	48·94
1	1·11	98·89	47	52·17	47·83
2	2·22	97·78	48	53·28	46·72
3	3·33	96·67	49	54·39	45·61
4	4·44	95·56	50	55·50	44·49
5	5·55	94·45	51	56·62	43·38
6	6·66	93·34	52	57·73	42·27
7	7·77	92·23	53	58·84	41·16
8	8·88	91·12	54	59·95	40·05
9	9·99	90·01	55	61·06	38·94
10	11·10	88·90	56	62·17	37·83
11	12·21	87·79	57	63·28	36·72
12	13·32	86·68	58	64·39	35·61
13	14·43	85·57	59	65·50	34·50
14	15·54	84·46	60	66·61	33·39
15	16·65	83·35	61	67·72	32·28
16	17·76	82·24	62	68·83	31·17
17	18·87	81·13	63	69·94	30·06
18	19·98	80·02	64	71·05	28·95
19	21·09	78·91	65	72·16	27·84
20	22·20	77·80	66	73·27	26·73
21	23·31	76·69	67	74·38	25·62
22	24·42	75·58	68	75·49	24·51
23	25·53	74·47	69	76·60	23·40
24	26·64	73·36	70	77·71	22·29
25	27·75	72·25	71	78·82	21·18
26	28·86	71·14	72	79·93	20·07
27	29·97	70·03	73	81·04	18·96
28	31·08	68·92	74	82·15	17·85
29	32·19	67·81	75	83·26	16·74
30	33·30	66·70	76	84·37	15·63
31	34·41	65·59	77	85·48	14·52
32	35·52	64·48	78	86·59	13·41
33	36·63	63·37	79	87·70	12·30
34	37·74	62·26	80	88·82	11·18
35	38·85	61·15	81	89·93	10·07
36	39·96	60·04	82	91·04	8·96
37	41·07	58·93	83	92·15	7·85
38	42·18	57·82	84	93·26	6·74
39	43·29	56·71	85	94·37	5·63
40	44·40	55·60	86	95·48	4·52
41	45·51	54·49	87	96·59	3·41
42	46·62	53·38	88	97·70	2·30
43	47·73	52·27	89	98·81	1·19
44	48·84	51·16	90·07	100	0
45	49·95	50·05			

Table XVI.

Sodium chloride.

S.G. *of solutions* (15·5° *C.*) *and per cent. by weight of* NaCl.

s.g.	Twaddell (approx.)	NaCl per cent.	s.g.	Twaddell (approx.)	NaCl per cent.
1·00725	1·4	1	1·11146	22·3	15
1·01450	2·9	2	1·11938	23·9	16
1·02174	4·3	3	1·12730	25·6	17
1·02899	5·8	4	1·13523	27·0	18
1·03624	7·2	5	1·14315	28·6	19
1·04366	8·7	6	1·15107	30·2	20
1·05108	10·2	7	1·15931	31·9	21
1·05851	11·7	8	1·16755	33·5	22
1·06593	13·2	9	1·17580	35·2	23
1·07335	14·7	10	1·18404	36·8	24
1·08097	16·2	11	1·19228	38·5	25
1·08859	17·7	12	1·20098	40·2	26
1·09622	19·2	13	1·20433	40·8	26·395
1·10384	20·7	14			

TABLE XVII.

Sulphuric acid. S.G. *of solutions and per cent. by weight of* H_2SO_4.

S.G. $\frac{15^\circ}{4^\circ}$ C. (in vacuo)	H_2SO_4 per cent.	S.G. $\frac{15^\circ}{4^\circ}$ C. (in vacuo)	H_2SO_4 per cent.	S.G. $\frac{15^\circ}{4^\circ}$ C. (in vacuo)	H_2SO_4 per cent.	S.G. $\frac{15^\circ}{4^\circ}$ C. (in vacuo)	H_2SO_4 per cent.
1·000	0·09	1·245	32·86	1·490	58·74	1·735	80·24
1·005	0·83	1·250	33·43	1·495	59·22	1·740	80·68
1·010	1·57	1·255	34·00	1·500	59·70	1·745	81·12
1·015	2·30	1·260	34·57	1·505	60·18	1·750	81·56
1·020	3·03	1·265	35·14	1·510	60·65	1·755	82·00
1·025	3·76	1·270	35·71	1·515	61·12	1·760	82·44
1·030	4·49	1·275	36·29	1·520	61·59	1·765	82·88
1·035	5·23	1·280	36·87	1·525	62·06	1·770	83·32
1·040	5·96	1·285	37·45	1·530	62·53	1·775	83·90
1·045	6·67	1·290	38·03	1·535	63·00	1·780	84·50
1·050	7·37	1·295	38·61	1·540	63·43	1·785	85·10
1·055	8·07	1·300	39·19	1·545	63·85	1·790	85·70
1·060	8·77	1·305	39·77	1·550	64·26	1·795	86·30
1·065	9·47	1·310	40·35	1·555	64·67	1·800	86·90
1·070	10·19	1·315	40·93	1·560	65·08	1·805	87·60
1·075	10·90	1·320	41·50	1·565	65·49	1·810	88·30
1·080	11·60	1·325	42·08	1·570	65·90	1·815	89·05
1·085	12·30	1·330	42·66	1·575	66·30	1·820	90·05
1·090	12·99	1·335	43·20	1·580	66·71	1·821	90·20
1·095	13·67	1·340	43·74	1·585	67·13	1·822	90·40
1·100	14·35	1·345	44·28	1·590	67·59	1·823	90·60
1·105	15·03	1·350	44·82	1·595	68·05	1·824	90·80
1·110	15·71	1·355	45·35	1·600	68·51	1·825	91·00
1·115	16·36	1·360	45·88	1·605	68·97	1·826	91·25
1·120	17·01	1·365	46·41	1·610	69·43	1·827	91·50
1·125	17·66	1·370	46·94	1·615	69·89	1·828	91·70
1·130	18·31	1·375	47·47	1·620	70·32	1·829	91·90
1·135	18·96	1·380	48·00	1·625	70·74	1·830	92·10
1·140	19·61	1·385	48·53	1·630	71·16	1·831	92·30
1·145	20·26	1·390	49·06	1·635	71·57	1·832	92·52
1·150	20·91	1·395	49·59	1·640	71·99	1·833	92·75
1·155	21·55	1·400	50·11	1·645	72·40	1·834	93·05
1·160	22·19	1·405	50·63	1·650	72·82	1·835	93·43
1·165	22·83	1·410	51·15	1·655	73·23	1·836	93·80
1·170	23·47	1·415	51·66	1·660	73·64	1·837	94·20
1·175	24·12	1·420	52·15	1·665	74·07	1·838	94·60
1·180	24·76	1·425	52·63	1·670	74·51	1·839	95·00
1·185	25·40	1·430	53·11	1·675	74·97	1·840	95·60
1·190	26·04	1·435	53·59	1·680	75·42	1·8405	95·95
1·195	26·68	1·440	54·07	1·685	75·86	1·8410	97·00
1·200	27·32	1·445	54·55	1·690	76·30	1·8415	97·70
1·205	27·95	1·450	55·03	1·695	76·73	1·8410	98·20
1·210	28·58	1·455	55·50	1·700	77·17	1·8405	98·52
1·215	29·21	1·460	55·97	1·705	77·60	1·8400	98·72
1·220	29·84	1·465	56·43	1·710	78·04	1·8395	98·77
1·225	30·48	1·470	56·90	1·715	78·48	1·8390	99·12
1·230	31·11	1·475	57·37	1·720	78·92	1·8385	99·31
1·235	31·70	1·480	57·83	1·725	79·36		
1·240	32·28	1·485	58·28	1·730	79·80		

TABLE XVIII.

Viscosities (*of some liquids*) $[\eta \times 100]$ *at* $40°$ *C.*

Water	0·657
Alcohol, methyl	0·450
,, ethyl...	0·827
Chloroform	0·465
Carbon tetrachloride	0·738
,, disulphide	0·319
Benzene	0·492
Turpentine	1·07
Hexane [C_6H_{14}]	0·264
Bromine	0·817
Aniline...	2·41
Formic acid	1·22
Acetic acid	0·90
Butyric acid	1·12
Sperm oil	17
Olive oil	34
Rape oil	40
Castor oil	248

APPENDIX I

A NEW TEST FOR OILS AND FATS

The authors have already pointed out (*Analyst*, 1918, **43**, 4—20) that the relative solubility of the natural oils and fats in a given solvent is a constant for each individual oil, providing that due allowance is made for free fatty acidity. They have been impressed with the practical utility of the test in the case of oils, and have found in the case of mixtures of oils that the resulting turbidity temperature of the mixture is virtually an arithmetic mean of the temperature figure for each of the component oils.

Since, in the case of soap and commercial fatty acid analysis, the original glycerides are not available, the authors investigated the turbidity temperatures obtained with the mixed fatty acids themselves with various solvents in order to discover whether an equal degree of discrimination were obtained as in the case of the oils.

They find the most suitable solvent for the purpose to be acetic acid of about 90 per cent. strength, and standardised as in the test for oils and fats[1] against pure Oleic acid or the mixed fatty acids of Almond Oil[2]. As was anticipated, the relation between the turbidity temperatures obtained with the mixed fatty acids corresponds closely with that of the oils themselves (pages 83—85), but the divergence is not quite so great between the lowest and the highest member, that is, the scale, in comparison with that of the oils, is somewhat compressed.

The presence of small amounts of undecomposed glycerides in the mixed fatty acids raises the turbidity temperature considerably; it is therefore essential that a complete saponification is obtained before the test is made, and the mixed fatty acids from soap, or as obtained commercially, should preferably be re-saponified by alcohol potash as in the titer test (page 42).

It is essential that the mixed fatty acids be thoroughly washed free from volatile acids, especially in the case of Coconut and Palm Kernel oils. Thus, the following results were obtained with samples of Coconut and Palm Kernel oils:

Procedure

20 grammes of oil taken.

Saponified with 100 c.c. of 2N alcoholic potash (90 °/₀ alcohol) under reflux for ½ an hour.

Alcohol evaporated off on water-bath.

100 c.c. of boiling water added to soap paste and heated until a perfectly clear solution resulted.

[1] See page 79.
[2] The two give almost identical results.

25 °/₀ H₂SO₄ added till in slight excess.

Fatty acids heated till perfectly clear and a clean separation obtained.

Acid liquor run off and fatty acids filtered through a dry filter without further treatment.

As so obtained and without washing, the turbidity values given by the mixed fatty acids were as follows:

Solvent = Acetic acid (80·0° C.   ⎰ 2 c.c. of fatty acids at 100° C.
    test to Oleic acid)   ⎱ 2 c.c. of Acetic acid at about 20° C.
                             (see page 79).

Coconut   = 21·5° C.
Palm Kernel = 30·0° C.

After thoroughly washing with hot distilled water until the washings showed no trace of acid, the washed acids were again filtered through a dry filter paper and the results obtained were as under:

Coconut   = 27·5° C.
Palm Kernel = 43·0° C.

In addition to its value in the analysis of soap and of commercial fatty acids, the test may be applied to oil analysis, conveniently in conjunction with the titer test, for which the same procedure is required. It has the advantage that no correction for acidity is necessary as in the case of oils, but, as mentioned before, care must be taken that a *thorough saponification* of the oil is obtained.

A further advantage offered by this method is that the result is not influenced by the presence of moisture in the oil, or in the acid, to anything like the same degree as in the case of the turbidity test with the oil. Thus, a rise of 1° C. in the turbidity temperature is produced by the addition of ·13 °/₀ of moisture to the solvent, the figure in the case of the stronger acid using the oils being ·029 °/₀ added moisture to produce an equal rise in the turbidity temperature (page 78).

The solvent has the additional advantage of not readily freezing at low temperatures and of being far less hygroscopic than in the case of the stronger acid.

We obtain the following results using the carefully washed mixed fatty acids from pure specimens of oils:

Turbidity Temperatures of Mixed Insoluble Fatty Acids

[Acetic acid = 80° C. test with oleic acid.]

The test is performed as on page 79.

Castor soluble	Cottonseed 76°	Lard 86°
Coconut 27·5°	Maize 76°	Arachis 87°
Palm Kernel 43°	Sesamé 77°	Tallow 83°
Linseed 69°	Chinese vegetable tallow 79°	Palm 83°
Soya bean 75°	Olive 78°	Rape 96°
Niger 75°	Almond 80°	Whale 72°

APPENDIX II

THE DETERMINATION OF GLYCERIN IN SOAP

As the glycerin normally retained by hard soaps and present in soft soaps made from fatty acids is usually small in amount, special precautions are necessary in order to obtain accurate results. There are two methods which the authors have found reliable and they give the preference to the second of these on account of its greater facility and speed. The first is, however, the recognised standard method and should be employed in case of dispute.

I

Procedure

1. Weigh out 100 grammes of soap and dissolve in hot water in a large beaker. Add 25 per cent. sulphuric acid until in slight excess.
2. Heat the beaker on the water-bath until the separated fatty acids form a clear and defined layer on the surface of the aqueous liquid.
3. Filter the whole liquid hot through a wet filter paper, washing acids well with hot water.
4. Add barium carbonate to the filtrate till in slight excess (test with litmus), then filter again.
5. Evaporate filtrate on water-bath until crystallisation begins, cool, treat the semi-solid residue with 95 °/₀ alcohol, and intermix thoroughly with a flat-sided stirring rod, crushing the lumps of crystal. Allow to stand for fifteen minutes.
6. Decant alcoholic solution and filter into a 200 c.c. beaker. Repeat lixiviation of the residue with alcohol several times, and evaporate off alcohol from combined washings by heating on a water-bath.
7. Make a mixture of 2 parts *absolute* alcohol and 1 part ether, and treat the syrupy residue several times with small quantities, filtering after each extraction into a tared, round-bottomed flask (see page 121).
8. Evaporate off the ether-alcohol from the flask rapidly, on the water-bath, taking care not to overheat the residual glycerin. Place the warm, dry flask in a vacuum desiccator over sulphuric acid or phosphoric oxide and exhaust with the air pump. Allow to remain overnight.
9. Weigh the flask. If the net weight of the glycerin comes within the prescribed amount, no further procedure is necessary.
10. If the glycerin is in excess or the actual percentage is required, determine its strength by the acetin method (page 188).

II

Procedure

1 —4 as in I.

5. Evaporate the filtrate on the water-bath to a bulk of about 100 c.c., cool, add ⅕ the quantity of silver carbonate stated on page 191 (D). Allow to stand for 15 minutes with occasional agitation, and then add excess of basic acid acetate solution (page 192, E).

6. Allow to stand for 15 minutes, filter and wash precipitate with a little distilled water. Evaporate filtrate to original bulk (100 c.c.), cool, add further quantity of basic lead acetate, and allow to stand, filter again.

7. Determine the glycerin present by the dichromate method (page 192).

SUBJECT INDEX

NAME INDEX

Printed in the United States
By Bookmasters